AUTOMOTIVE COMPUTER CONTROL:
Emission and Tune-Up

AUTOMOTIVE COMPUTER CONTROL: Emission and Tune-up

PRENTICE-HALL, Englewood Cliffs, New Jersey 07632

Library of Congress Cataloging-in-Publication Data

Main entry under title:

Automotive computer control.
    Includes index.
      1. Automobiles—Electronic equipment. 2. Automobiles
—Electronic equipment—Maintenance and repair.
I. Tune-up Manufacturers Institute.
TL272.5.A9535 1986   629.2'549   85-28141
ISBN 0-13-054727-1
ISBN 0-13-054701-8 (pbk.)

TMI STAFF: Ralph W. Van Demark and Diane Johnson
COORDINATING EDITOR: Bob Skakun
Editorial/production supervision and
  interior design: Tom Aloisi
Cover design: 20/20 Services, Inc.
Manufacturing buyer: Rhett Conklin

© 1986 by Tune-Up Manufacturers Institute,
Teaneck, New Jersey 07666

Portions of this book were previously published under the
title, *Automotive Emission Control and Tune-Up Procedures, 3rd Ed.*,
by the Ignition Manufacturers Institute.

All rights reserved. No part of this book may be
reproduced, in any form or by any means,
without permission in writing from the Tune-Up Manufacturers Institute
and the publisher.

The information and instructions contained in this text
have been prepared by experienced mechanics and engineers.
However, the author cannot and does not guarantee the accuracy
of this information or the efficacy of these instructions.
The author expressly disclaims any responsibility for damages
arising from use of or reliance on the information and
instructions contained herein.

Printed in the United States of America

10 9 8 7 6 5 4 3 2 1

ISBN: 0-13-054727-1
ISBN: 0-13-054701-8 {PBK}   025

Prentice-Hall International, Inc., *London*
Prentice-Hall of Australia Pty. Limited, *Sydney*
Editora Prentice-Hall do Brasil, Ltda., *Rio de Janeiro*
Prentice-Hall Canada Inc., *Toronto*
Prentice-Hall Hispanoamericana, S. A., *Mexico*
Prentice-Hall of India Private Limited, *New Delhi*
Prentice-Hall of Japan, Inc., *Tokyo*
Prentice-Hall of Southeast Asia Pte. Ltd., *Singapore*
Whitehall Books Limited, *Wellington, New Zealand*

CONTENTS

The Tune-up Manufacturers Institute, xi
Members of the Tune-up Manufacturers Institute, xii
Introduction to Clean Air, xv
Introduction to Text, xvii

1 GENERAL INFORMATION, 1

Diagnosis and Testing, 3
Testing Sections, 5
Testing Essentials, 7
Four-stroke Cycle, 9

2 MAGNETISM AND ELECTRICITY, 11

Electrical Terms, 13
OHM's Law, 15
Electrical Circuits, 17
Magnetism, 21
Electromagnetic Fields, 23
Automotive Electrical System, 25
Automotive Ground Circuits, 27
Wiring and Schematic Diagrams, 29
Circuit Testing Tools, 31
Introduction to Solid-state Electronics, 35

3 BATTERY, 39

Battery Visual Checks, 41
Battery Leakage Test, 43
Battery Specific Gravity Test, 45
Temperature-corrected Hydrometer, 47
Battery Capacity (Load) and Three-minute Charge Tests, 49

4 STARTING SYSTEM, 53

Starting Circuit, 55
The Starting Motor, 57

© Copyright 1986, Tune-Up Manufacturers Institute

Starting Motor Drives, 59
Starting Switch, 61
Starting Motor Amperage Draw Test, 63
Starter Insulated Circuit Test, 65
Starter Ground Circuit Test, 67

5 CHARGING SYSTEM, 69

Constant Voltage Charging System, 71
Alternator Operating Principles, 73
Alternator Components, 75
Diodes, 77
Transistors, 79
Alternators with a Diode Trio, 81
AC Charging System (with Ammeter), 83
AC Charging System (with Indicator Lamp), 85
AC Charging Circuit Indicator Lamps, 87
Alternator Voltage Control, 89
Mechanical Voltage Regulator, 91
Mechanical Voltage Regulator with Indicator Lamp, 93
Electronic (Transistor) Voltage Regulator, 95
Test for Alternators with Mechanical Regulators, 97
Test for Alternators with Electronic Regulators, 99
Alternator Testing—General Motors Delcotron, 101
Alternator Testing—General Motors Delcotron with Integral Regulator, 105
Alternator Testing—Chrysler Corporation I, 107
Alternator Testing—Chrysler Corporation II, 111
Alternator Testing—Ford Motor Company, 113
Alternator Testing Factors, 115
Diode Tests, 119
Diode Trio, 121
Field Winding Tests, 123
Stator Winding Tests, 125
Alternator Charging System Service Precautions, 127
Generator Operating Principles, 131
Generator Testing, 133

6 THE IGNITION SYSTEM, 135

Ignition Circuit (Breaker-point Type), 137
Ignition System, 139
Distributor Assembly, 141
Dwell Angle, 143
Spark Advance Timing, 147
Centrifugal Advance Mechanism, 149
Spark Advance, 151

Ignition Coil, 153
Coil Polarity, 157
Available Voltage, 159
Condenser Construction, 161
Condenser Action, 163
Distributor Cap and Rotor, 167
Secondary Circuit Suppression, 169
Cylinder-numbering Sequence and Firing Orders, 173
Spark-plug Heat Range, 175
Spark-plug Features, 177
Capacitive Discharge Ignition System, 181

7 ELECTRONIC IGNITION SYSTEMS, 183

Chrysler Electronic Ignition System (Except Lean Burn), 185
Chrysler Electronic Lean Burn System, 187
Troubleshooting the Chrysler Electronic Ignition System, 191
Delco High-energy Ignition (HEI) System, 195
Troubleshooting the Delco High-energy Ignition (HEI) System, 197
Ford Solid-state Ignition System (Pre-1977, Dura-Spark I, II), 201
Troubleshooting the Ford Solid-state Ignition System (Pre-1977), 203
Troubleshooting the Ford Solid-state 1977 Ignition System (Dura-Spark Systems I, II), 205
Prestolite Breakerless Inductive Discharge Ignition System, 207
Troubleshooting the Prestolite BID Ignition System, 209
Ford Dual-mode Timing Ignition Systems (1978–1980), 211
Thick-film Integrated (TFI) Systems, 213
Hall-Effect Switch, 215
Detonation Sensors, 217
General Motors Electronic Spark Timing (EST CCC, 1980-on), 221

8 FUEL SYSTEMS, 225

Fuel Systems, 227
Fuel Pump, 231
Electric Fuel Pumps, 233
Float System, 237
Idle Circuit, 239
Main Metering System, 241
Vacuum-controlled Metering Rod, 243
Power System, 245
Accelerating System, 247
Choke, 249
Automatic Choke, 251
Electric Choke-assist Systems, 255
Electronic Carburetors, 257
Carburetor Adjustment for Nonexhaust Emission Controlled Engines, 259

© Copyright 1986, Tune-Up Manufacturers Institute

9 EMISSION CONTROL SYSTEMS, 261

Introduction to Vehicle Emission Control Systems, 263
Exhaust Emission Control Systems, 267
Air Injection Systems, 269
Heated Carburetor Air Systems, 273
Testing Heated Carburetor Air Systems, 277
Idle Stop Solenoids, 281
Hot Idle Compensator Valve, 285
Deceleration (Decel) Valve, 287
Exhaust Gas Recirculation (EGR) Systems, 289
Dual Diaphragm EGR Valve, 293
Backpressure Transduced EGR Valves, 295
Computerized EGR Systems, 299
General Motors EGR Control Systems, 301
Late-model Ford EGR Systems, 305
Testing EGR Systems, 309
Fuel Evaporation Emission Control Systems, 313
Late-model Purge Control, 317
Manifold Heat Control Valve (Heat Riser), 321
Electric Grid Early Fuel Evaporation (EFE), 323
Positive Crankcase Ventilating Systems, 325
Positive Crankcase Ventilation System Tests, 329
Catalytic Converters, 333
Oxygen Sensor, 337
Transmission-Regulated Spark (TRS) System, 341
Spark-delay Valve, 343
Exhaust Emission Control System Assist Units, 345
Testing Exhaust Emission Control System Assist Units, 349
Carburetor Adjustment for Exhaust Emission Controlled Engines, 355

10 COMPUTERIZED ENGINE CONTROL SYSTEMS, 359

Introduction to Computerized Engine Control Systems, 361
General Motors Computer Command Control (CCC), 363
General Motors CCC System Diagnostic Codes, 369
Ford Electronic Engine Control (EEC), 373
Ford EEC System Self-test Capabilities, 379
Chrysler Electronic Fuel Control (EFC) System, 381

11 ELECTRONIC FUEL INJECTION, 385

Introduction to Electronic Fuel Injection, 387
Multipoint Electronic Fuel Injection, 389
Throttle Body Fuel Injection (TBI), 393

12 TMI TUNE-UP PROCEDURE, 395

TMI Tune-up Procedure, 397
Cranking Voltage Test, 399
Charging Voltage Test, 401
Power Loss, 403
Compression Test, 405
Spark Plugs, 409
Ignition Timing, 411
Supplementary Tune-up services, 419

GLOSSARY

Electrical and Electronic Symbols, 421
Abbreviations and Glossary of Terms, 429

INDEX, 437

© Copyright 1986, Tune-Up Manufacturers Institute

The TUNE-UP MANUFACTURERS INSTITUTE

The TUNE-UP MANUFACTURERS INSTITUTE, through the combined resources, knowledge, experience and service abilities of its independent tune-up parts manufacturing members, provides the following for the independent automotive parts distributors and their service shop customers:

1. A permanent and profitable business

2. Quality precision-built parts

3. Competitively and profitably priced parts

4. Immediate availability of parts through wide distribution channels

5. Complete educational programs

© Copyright 1986, Tune-Up Manufacturers Institute

THE TUNE-UP MANUFACTURERS INSTITUTE

The Tune-up Manufacturers Institute is composed of a group of independent automotive parts manufacturers who have organized to effectively serve the automotive parts distributors, service shops, and the motoring public.

The basic function of the Institute is to assist automotive parts distributors and service shops to operate effectively and profitably by making available quality precision-manufactured parts. This, in turn, allows them to offer their customers prompt and satisfactory service.

The Tune-up Manufacturers Institute, through the combined resources, knowledge, experience, and service abilities of its members, provides the following benefits for parts distributors and their service shop customers:

1. A permanent and profitable business.
2. Quality, precision-built parts.
3. Competitively and profitably priced parts.
4. Immediate availability of parts through wide distribution channels.
5. Educational programs.

The Tune-up Manufacturers Institute and its members have the sincere desire to make available premium-quality automotive parts, constant service, technical publications, service manuals, and an automotive tune-up course—all of which have been developed to assist the automotive serviceman to operate a successful and profitable business and to help raise his vocational standard to the level of other professional tradesmen.

© Copyright 1986, Tune-Up Manufacturers Institute

Members of the
TUNE-UP MANUFACTURERS INSTITUTE
222 Cedar Lane
Teaneck, New Jersey 07666

Autolite Division
Providence, RI 02916

Autolite Division
Allied Automotive
105 Pawtucket Avenue
East Providence, Rhode Island 02916

BWD Automotive Corporation
11045 Gage Avenue
Franklin Park, Illinois 60131

Champion Spark Plug Company
P.O. Box 910
Toledo, Ohio 43661

Echlin Manufacturing Company
P.O. Box 472
Echlin Road and U.S. 1
Branford, Connecticut 06405

Filko Automotive Products
Division of F & B Manufacturing Company
5480 N. Northwest Highway
Chicago, Illinois 60630

General Automotive Specialty Co., Inc.
U.S. 1 and 130
P.O. Box 3042
North Brunswick, New Jersey 08902

© Copyright 1986, Tune-Up Manufacturers Institute

Guaranteed Parts
A Division of Gulf & Western Manufacturing Company
Auburn Road
Seneca Falls, New York 13148

Kem Manufacturing Co., Inc.
River Road and Maple Avenue
Fair Lawn, New Jersey 07410

Tom McGuane Industries, Inc.
F & E Manufacturing Division
32031 Townley Avenue
Madison Heights, Michigan 48071

Sorensen Industries, Inc.
1115 Cleveland Avenue
Glasgow, Kentucky 42141

Standard Motor Products, Inc.
37–18 Northern Boulevard
Long Island City, New York 11101

Switches, Inc.
6131 W. 80 St.
Indianapolis, Indiana 46278

Niehoff Automotive Parts

TRW, Inc.
Replacement Parts Division
8001 East Pleasant Valley Road
Cleveland, Ohio 44131

© Copyright 1986, Tune-Up Manufacturers Institute

United Technologies
Automotive, Inc.
5200 Auto Club Drive
Dearborn, Michigan 48126

Valley Forge Products
Division of Avnet, Inc.
150 Roger Avenue
Inwood, New York 11696

Wells Manufacturing Corp.
2-26 South Brooke Street
P.O. Box 70
Fond du Lac, Wisconsin 54935

© Copyright 1986, Tune-Up Manufacturers Institute

INTRODUCTION TO CLEAN AIR

Emission control systems and components have been a part of the automobile for almost 20 years. When scientists determined that the quality of America's air had reached a life-threatening level, the automobile was designated a primary contributor of airborne pollutants. Vehicle emission standards tightened every year, requiring manufacturers to reduce the output of hydrocarbons (HC), carbon monoxide (CO) and oxides of nitrogen (NO_x). With the stricter regulations of each succeeding year, the mechanic was also forced to keep pace with a flood of new and different technologies. Some mechanics accepted the clean-air challenge and became successful emission control technicians. Others approached the subject with a negative attitude and have found their ability to service the modern automobile severely limited.

The pursuit of clean air through legislation has not only affected the complexity of the vehicle, it has also spawned a series of laws affecting the mechanic and driving public. The manufacturer is required to grant a warranty of five years or 50,000 miles on all emission-control-related devices. The vehicle owner is required to maintain his or her car in accordance with state inspection laws and to purchase higher-priced unleaded gasoline. Federal EPA regulations are also requiring the states to enact stricter inspection procedures. For example, July 1, 1985, federal standards reduced hydrocarbon and carbon monoxide limits for vehicles constructed after 1980. These standards changed from a limit of 3.0 percent CO and 300 ppm hydrocarbons to 2.0 percent CO and 200 ppm hydrocarbons. The burden of enforcement has been placed on the state, which must implement Inspection/Maintenance (I/M) programs. Some states operate government-run inspection stations, while others license independent garages to service the public. Regardless of the method used to enforce clean-air laws, most states are beginning to push for emissions certification testing of all mechanics. Whether you agree or disagree with the federal mandates, a thorough working knowledge of emission control systems is a necessity for the mechanic.

Fuel Economy Requirements

During the winter of 1973–1974, America experienced its first fuel shortage since World War II. Five years later, a second and more severe fuel crisis occurred. These crises caused fuel prices to soar, and energy conservation became a national issue. Car makers scrambled to produce small, efficient, and economical automobiles. The Department of Energy was formed and new federal regulations came into existence. Corporate average fuel economy (CAFE) requirements forced manufacturers to sell fuel-efficient cars.

CAFE requirements and clean-air regulations were at odds with each other. Past emission control methods reduced fuel economy and hampered performance. New technologies had to be found to meet the social needs for clean air and reduced energy consumption. The world of high technology held the answers to reduced emissions, fuel economy, and improved performance. The computer has been able to solve these problems and promises to improve other aspects of automobile operation, such as braking performance and comfort controls.

Technology is advancing at an extraordinary rate and the technician must keep pace. The complexity of the computer-controlled automobile requires the mechanic to continually study and update his or her skills. These new systems do not tolerate tampering or disconnection, nor does the government. Antitampering laws allow the imposing of up to $10,000 fines on technicians caught removing or modifying any emissions-related components. Roadside spot checks are being used to keep everyone honest.

The car owner has met his responsibilities when he brings his car to the garage. Responsibility then rests with the tune-up specialist. The technician must be qualified to tune the engine and test emission control systems. The technician qualifies himself by keeping abreast of technological advances and modern test equipment. Some mechanics incorrectly assume that new tools and sophisticated testers will fix the car. Remember, a piece of equipment is only as good as the mechanic using it.

The customer expects his or her car to pass inspection and deliver good fuel economy. The technician that satisfies these expectations will enjoy a fruitful career. This TMI textbook is emission-control and computer-oriented to assist the engine performance technician in acquiring the knowledge to satisfy both the customer's needs and government regulations.

INTRODUCTION TO TEXT

Prior to 1970, all vehicles were very similar with regard to ignition and fuel systems. The Kettering ignition (point type) and carburetor were the standard for virtually all domestic and import cars. Due to the environmental and economic challenges of the past 15 years, the automobile has undergone extensive changes. The points and condenser disappeared 10 years ago, and the distributor itself is in the process of being eliminated. The carburetor has become a more sophisticated electronic device and is rapidly being replaced by electronic fuel injection.

In the process of meeting the demands of environmental protection regulations, vehicle manufacturers developed hundreds of different emission control components and systems. Some of these systems proved beneficial and reliable, while others did not live up to expectations and were either revised or replaced.

During the 1970s, America had its first experiences with peace-time fuel shortages. The rush toward improved fuel economy was on and the government imposed new fuel economy regulations on the vehicle manufacturers. These regulations are known as corporate average fuel economy (CAFE) requirements. In striving for improved fuel economy, engineers and service technicians found that clean-air technology and fuel economy were at odds with each other. Methods used to reduce emissions also seemed to reduce fuel economy. A new direction had to be taken.

In the 1980s, the computer has found its way into virtually every automobile sold in America. While the computer had been used on imports as early as 1968 and on various domestic cars during the late 1970s, its commands were limited to fuel or ignition timing control. With the development of digital techniques, the computer was ready to expand its sensory and command capabilities. Fuel economy, emission controls, and engine performance (drivability) are now all monitored and controlled by one or more logic-type computers. These technical advances have allowed the modern automobile to meet the strict emission and fuel economy regulations, and at the same time they improve engine performance that had been lost in the pursuit of clean air.

There is no questioning the fact that these technical advances have improved the overall performance of the vehicle, but other problems have arisen. The technical sophistication of these systems has increased at a fantastic rate, placing a severe strain on the mechanic's ability to keep pace. The mechanic must again become a student to understand the intricacies of the new automobile, as well as go back and relearn those things he or she did not pay close attention to in the past.

This book is not intended to be a repair manual. The number of different ignition, fuel, and emission control systems prevents the Tune-up Manufacturers Institute from describing them all, let alone detailing the diagnostic and repair procedures for each system. The use of individual repair manuals is an absolute necessity in the repair of an automobile. This text is intended to give the student an understanding of how these systems function on the modern vehicle. It is nearly impossible for a mechanic to repair something if he or she does not know how it works. The use of this text in conjunction with the vehicle repair manual can simplify testing and diagnostic procedures and lead to a satisfactorily completed repair.

AUTOMOTIVE COMPUTER CONTROL:
Emission and Tune-Up

1
GENERAL INFORMATION

Notes

DIAGNOSIS AND TESTING

There are two steps involved in the repair of the modern automobile: diagnosis and testing. Many mechanics attempt to solve a problem by replacing parts until they have found the cure. Sometimes the problem is found by replacing only a few components, but most often this process yields only mediocre results with greatly increased cost to the customer. Parts replacing also detracts from the mechanic's ability to properly deal with future problems.

Diagnosis is the identification of the problem area. Since this text deals with engine performance, there are five systems to be concerned with: engine mechanical, ignition, fuel, starting, and charging. To identify the faulty system, always begin with the driver's complaint. By listening to the customer, a mechanic can often diagnose which system requires attention or if the vehicle operator is creating the problem. The latter is often the most difficult to solve.

The diagnostic process begins with finding the answers to a series of questions. As an example, consider the typical "no start" complaint. Does the engine turn over properly when the key is turned? If the answer to this question is yes, the starting system could be eliminated from the list of possibilities. Is the proper amount of fuel being supplied? Is the ignition system providing adequate spark at the proper time? Does the engine generate sufficient pressure in the cylinders (compression)? All these questions must be asked, and the one with the negative answer will identify the system requiring attention.

Testing is an examination of the components within a system. The individual parts of a system are the links of a chain. The operation of that chain will only be complete if all the links are interconnected properly. The actual testing procedures range from simple inspection of wires, connectors, and hoses to the intense study of an oscilloscope pattern. Regardless of the equipment and procedures available, an intelligent and systematic approach must be used. Begin with the most obvious possibilities first. Many a mechanic has spent hours testing a fuel system, only to find that the vehicle had no fuel in the tank. Do not presuppose that the most complex component is at fault, because those complicated pieces usually require the most time-consuming test procedures. Do not start testing in the middle of a system. In the case of an ignition problem, check for the availability of voltage to the system, and do not automatically blame the control module.

The ability to properly diagnose and test is dependent on the mechanic's knowledge and understanding of the automobile's operation. Repair of the modern automobile requires more mental power than physical strength.

© Copyright 1986, Tune-Up Manufacturers Institute

Figure No. 1

TESTING SECTIONS

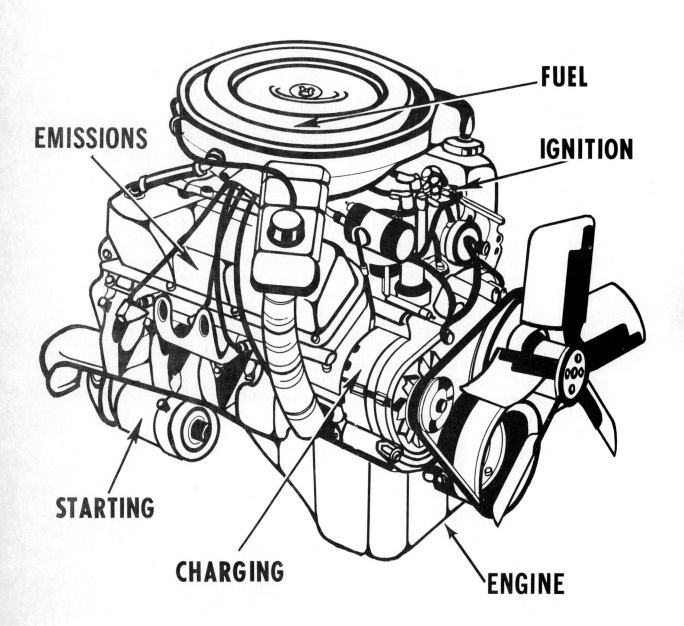

TESTING SECTIONS

There are five important engine areas that must function properly in order to insure good performance, economy, and dependability—the three factors in which practically every motorist is interested. The five areas to be considered are the engine, the starting system, the charging system, the ignition system, and the fuel system. Only when these five areas are restored to normal operating condition can a satisfied customer be developed and greater profits realized. These five areas, as they influence tune-up, will be covered in this course.

The emission control systems and assist devices designed into the modern automobile are *not* systems separate from the five testing areas. These systems and units are part of the distributor, ignition system, ignition timing, carburetor, and fuel system and must be tested and serviced at the same time the basic system is being diagnosed and tuned. This is the only way the desired results of a tuned engine and limited emissions can both be achieved at the same time.

Much of this course will be devoted to a study of the electrical system due to the great part it plays in the overall function and performance of the automobile. Almost every important part or operation of the automobile depends to some degree on the proper functioning of the electrical system.

The electrical accessories on late-model automobiles are many in number compared to early vehicles. Electrical systems on today's vehicles are being called upon to perform functions beyond those required a few years ago. Although service operations on the modern automobile sometimes seem difficult, the proper understanding of the units in the electrical system and their function converts these tasks into quickly accomplished profitable service jobs.

This course has been developed to assist you by increasing your understanding of these tune-up service operations.

© Copyright 1986, Tune-Up Manufacturers Institute

Figure No. 2

TESTING ESSENTIALS

The five elements of quality automotive tune-up are:

1 - Trained persons
2 - Dependable test equipment
3 - Accurate specifications
4 - Specific test procedure
5 - Quality replacement parts

© Copyright 1986, Tune-Up Manufacturers Institute

TESTING ESSENTIALS

There are *five* basic elements needed to perform a quality tune-up.

THE FIRST ESSENTIAL IS A TRAINED SERVICEPERSON. The success of a tune-up business is particularly dependent on this factor. A trained personnel is an observant serviceperson who by virtue of his/her knowledge, ability, and experience is able to locate the trouble and to restore the vehicle to its original operating condition quickly and efficiently. Since he/she is a trained individual, he/she knows the value of, and gets the maximum results from, the test equipment he/she uses. This is true because test equipment is only as good as the person who operates it.

THE SECOND ESSENTIAL IS DEPENDABLE TEST EQUIPMENT. The use of dependable test equipment is an extension of the serviceperson's senses, permitting him to locate trouble quickly and accurately. The increasing complexity of the modern automobile engine makes the use of test equipment a necessity. Several members of the Tune-Up Manufacturers Institute merchandise a line of dependable test equipment.

THE THIRD ESSENTIAL IS ACCURATE SPECIFICATIONS. Accurate specifications serve as a standard to determine the service tolerances of the vehicle. By comparing the readings obtained from the test equipment with the specifications, the trained serviceperson can quickly determine the service necessary to restore the vehicle to an efficient operating condition. These specifications are compiled by your Tune-Up Manufacturers Institute supplier.

THE FOURTH ESSENTIAL IS A SPECIFIC TEST PROCEDURE. The test procedure specified in this course contains the elements necessary to cover the greatest majority of engine and accessory malfunctions. Further, the steps of this procedure are arranged in a sequence that permit the tests to be easily conducted, in a logical order, with a minimum expenditure of time.

THE FIFTH ESSENTIAL, EQUAL IN IMPORTANCE TO ANY OTHER, IS QUALITY REPLACEMENT PARTS. The best assurance a tune-up specialist has of performing a tune-up that will keep an engine operating efficiently for thousands of miles is the use of quality replacement parts. The Tune-Up Manufacturers Institute organization sponsoring this training course features premium quality parts.

Remember—your tune-up business can only be as good as the service you render and the products you use.

Figure No. 3

FOUR STROKE CYCLE

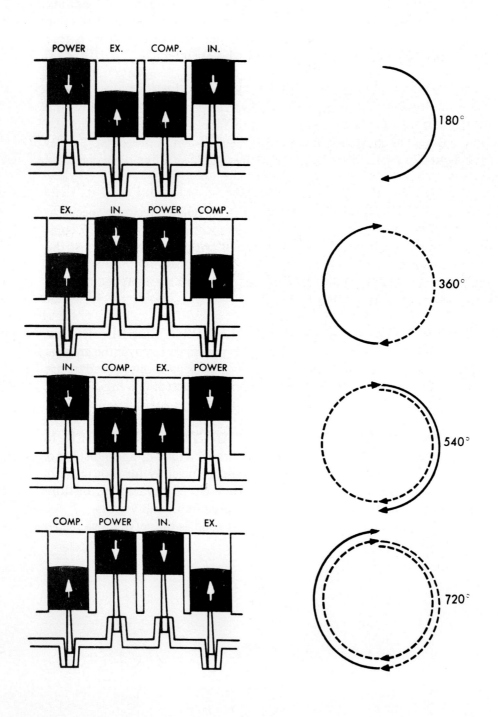

FOUR-STROKE CYCLE

The gasoline engine is an internal combustion heat engine that develops its power by burning a mixture of gasoline and air in the engine's cylinders. This engine is designed to operate on a four-stroke cycle principle. Each stroke is a movement of a piston, either upward or downward, in its cylinder.

The strokes occur in the following sequence: intake, compression, power, and exhaust.

The intake stroke (downward) serves to draw the air-fuel mixture from the carburetor into the cylinder through the open intake valve. The compression stroke (upward) compresses the mixture to approximately 150 pounds per square inch, a pressure at which the gaseous mixture is suitable for efficient combustion. The power stroke (downward) is the result of the spark plug igniting the compressed fuel charge. The rapid burning of the fuel produces tremendous heat and pressure, which expands the gases and raises the combustion pressure to more than four times the compression pressure. This pressure, exerted on the head of the piston, forces the piston down, producing the power stroke. The piston, on the return exhaust stroke (upward), forces the burned gases through the open exhaust valve, out of the cylinder, and into the exhaust system. With the cylinder cleansed of the exhaust gases, the cylinder is ready for another intake stroke and a repetition of the four-stroke cycle. The cycle of strokes, in the sequence listed, continues as long as the engine is in operation.

Theoretically, each stroke lasts for 180 degrees of crankshaft rotation. In actual practice, however, the length of these strokes is modified somewhat for better engine performance and efficiency.

All the cylinders in an engine are fired in every 720 degrees of crankshaft rotation regardless of the number of cylinders in the engine. The more cylinders an engine has, the more power strokes there will be in these two crankshaft revolutions, increasing the engine's power output and performance.

Notes

2
MAGNETISM AND ELECTRICITY

Figure No. 4

COMMONLY USED ELECTRICAL TERMS

CIRCUIT
CONDUCTOR
INSULATOR
AMPERE
VOLT
OHM
WATT

© Copyright 1986, Tune-Up Manufacturers Institute

ELECTRICAL TERMS

To expertly diagnose electrical system troubles, the tune-up specialist must understand the electrical terms commonly used. So that a student can readily grasp the meaning of these terms, a water analogy is used, comparing the movement of electricity through a wire to the flow of water through a pipe.

A *circuit* is a path through which current can flow. Current flows through a circuit much like water flows through a pipe. The principal requirement of any circuit is that it must form a complete path. In tracing circuits, it is important to start at the source of electric power, either the battery or the alternator, then follow the path of current flow through the components of the insulated circuit, and return to the source through the ground circuit. A circuit is *not* complete if the current cannot return to its source.

A *conductor* is a material that will pass electrical current efficiently, just as a clean pipe is a good conductor for water. The ability of a conductor to carry current not only depends upon the material used but also on its length, its cross-sectional area, and its temperature. A short conductor offers less resistance to current flow than a long conductor. A conductor with a large cross section will allow current to flow with less resistance than a conductor with a small cross section. For most materials, the higher the temperature of the material, the more resistance it offers to the flow of electrical current.

An *insulator* is a material that will not pass current readily. An insulator is used to prevent leakage of electrical current.

An *ampere* is a unit of measurement for the flow of a quantity of electrical current. In terms of water analogy, this would be compared to gallons.

A *volt* is a unit of measurement of electrical pressure, or electromotive force. Voltage is sometimes described as a difference of potential between the positive and negative terminals of a battery or generator. In terms of the water analogy, this pressure would be compared to pounds per square inch. For current to flow through a circuit, voltage must be applied to the circuit.

An *ohm* is a unit of electrical resistance opposing current flow. Resistance varies in different materials and varies with temperature. In terms of the water analogy, this resistance would be compared to a restriction in a pipe.

A *watt* is a unit of electrical power and is obtained by multiplying volts and amperes. As a point of interest, 746 watts is equal to one mechanical horsepower.

© Copyright 1986, Tune-Up Manufacturers Institute

OHM'S LAW

<u>1 VOLT</u> IS NECESSARY TO PUSH <u>1 AMPERE</u> THROUGH <u>1 OHM</u> OF RESISTANCE.

$$AMPERES = \frac{VOLTS}{OHMS}$$

$$VOLTS = AMPERES \times OHMS$$

$$OHMS = \frac{VOLTS}{AMPERES}$$

OHM'S LAW

Ohm's law, a basic electrical rule, states that 1 volt (of pressure) is required to push 1 ampere (of current) through 1 ohm (of resistance).

This fundamental rule is applicable to all electrical systems and is of outstanding importance in understanding electrical circuits. It is used in circuits and parts of circuits to find the unknown quantity of voltage, current, or resistance when the other two quantities are known.

Using Ohm's law, the unknown quantity is determined as follows:

1. To find the amperes—divide the voltage by the resistance.
2. To find the voltage—multiply the amperes by the resistance.
3. To find the resistance—divide the voltage by the amperage.

Remember—the current that flows in an electrical circuit is the balance between the applied voltage and the total circuit resistance.

It will not be necessary for you to stop and compute electrical values, using Ohm's law, during a tune-up. It is advisable, however, that you have a basic understanding of its application. Your test equipment works out these problems for you, giving you the answers in the form of meter indications. With the assistance of the equipment, your attention is quickly directed to the source of the trouble.

As a general automotive electrical system troubleshooting rule, remember— if the voltage remains constant, as it usually does except in the case of a discharged battery, an increase or decrease in current flow can only be caused by a change in resistance.

ELECTRICAL CIRCUITS

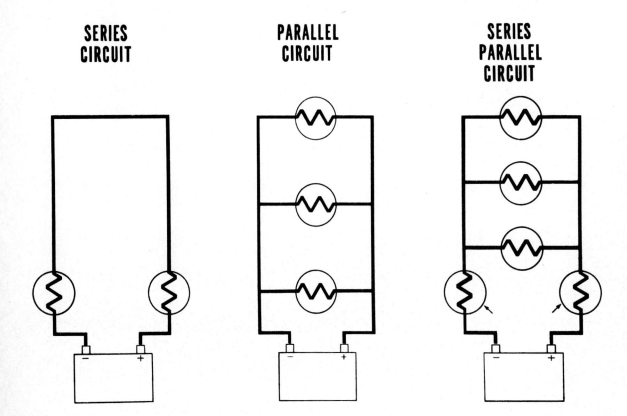

ELECTRICAL CIRCUITS

The symbol E represents *electromotive force* (electron-moving force), commonly referred to as *volts* or electrical pressure.

The symbol I represents *intensity* or current flow in *amperes*.

The symbol R represents *resistance*, which is measured in *ohms*. The symbol Ω for omega, the last letter in the Greek alphabet, is used as the ohm symbol, to avoid using the letter O, which can easily be mistaken for the numeral 0 (zero).

In summation, E is volts, I is amperes, and R is ohms.

The automotive electrical system is a combination of interrelated circuits. Many of the electrical components in a system have self-contained circuits. Diagnosing trouble will require a knowledge of where to look when certain conditions are indicated. The ability to trace a circuit will be of great value in pinpointing the difficulty.

Series Circuit

A series circuit is a circuit where there is only one path in which the current can flow. Any number of lamps, resistors, or other devices having resistance can be used to form a series circuit. The total resistance of a series circuit is the sum of the individual unit resistances. The more resistances that are added to the circuit, the higher will be the total resistance. Since there is only one path for current to flow in a series circuit, this means that all current must pass through each resistance in the circuit. If an opening occurs in any portion of a series circuit, the circuit will become inoperative. This will result in an incomplete circuit. A good illustration of this circuit is the old-style Christmas tree lights. If one bulb burns out, it opens the circuit and the rest of the bulbs go out.

The current flow in a series circuit is controlled by the total resistance of the circuit and the voltage applied. The current flow (amperes) will be the same in all places in the circuit. If two ammeters are connected in different places in the series circuit, both ammeters will read alike. If more resistance is added to the circuit, the amperage will become less, and if resistance is removed from the circuit, the amperage will increase.

As voltage moves current through a resistor, some of the force is expended, resulting in a loss or drop in voltage. This *voltage drop* always accompanies current moving through a resistance. Therefore, in a series circuit, the total voltage will always equal the sum of the voltage drop across the individual resistance units. The total voltage or the voltage across each resistance can be measured with a voltmeter, and this method, called *voltage drop test,* is widely used to determine circuit conditions.

Parallel Circuit

The circuit that has *more than one* path for current is called a parallel circuit. Parallel resistances connected across a voltage source have the same voltage applied to each resistance. The resistance of the individual units may or may not be the same value. Since the current divides among the various branches of the circuit, the current through each branch will vary, depending upon the resistance of the branch. However, the total current flow will always equal the sum of the current in the branches. The total resistance of a parallel circuit is always less than the smallest resistance in the circuit. If a break occurs in a parallel circuit, the circuit is not rendered inoperative, because there is more than one path for current to flow back to its source. An illustration of this is street lights. If one bulb burns out, the others remain lit.

An important thing to remember in a parallel circuit is that the voltage applied remains constant at each branch.

Series-Parallel Circuit

Many practical applications in the electrical system of the automobile depend upon a combination of series circuit and parallel circuit. This is called a *series-parallel circuit*. Such combinations are frequently used, particularly in electric motors and control circuits.

Notes

MAGNETISM & PERMANENT MAGNETS

MAGNETIC FIELD

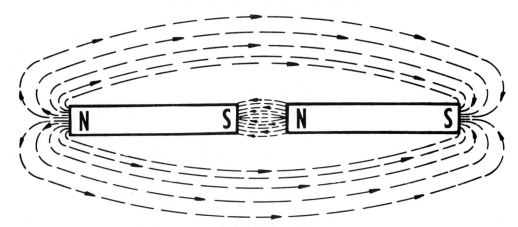

UNLIKE POLES ATTRACT

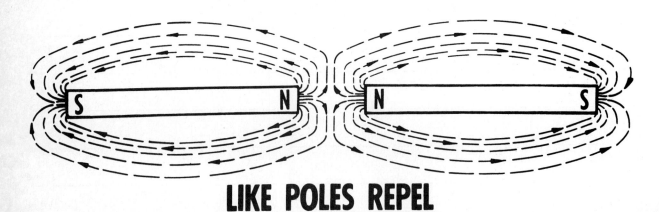

LIKE POLES REPEL

MAGNETISM

Magnetism is essential to the operation of electric motors, generators, solenoids, and most of the new sensory devices found on the automobile. Magnetism may be considered the mechanical form of electrical power. To fully comprehend the function of the vehicle's electrical systems, it is necessary to be familiar with the laws of magnetism.

Basic Theory of Magnetism

Magnetism is an invisible force that attracts materials containing iron. When an iron particle comes under the influence of a magnet, it is said to have entered the magnetic field. The *magnetic field* is the area around the magnet containing lines of force. These lines of force that compose the field are referred to as *flux*. Strong magnets will have many flux lines, thus creating larger and more intense magnetic fields than weaker magnets.

Magnets have a north and a south pole. The flux lines travel from north to south poles through the magnetic field. The lines of force move from south to north poles within the magnet. Remember that these lines are imaginary and are only used to create a graphic illustration of magnetic properties.

The principles of polarity become evident when two magnets are placed next to each other. When opposite poles are placed together, the magnets are attracted to one another. The two magnets are drawn together by the actions of their combined fields creating a single magnet. If like poles of the magnets are placed together they will repel each other. This happens because the lines of force are in opposition and the magnetic field cannot be combined. The polarity of magnetic fields makes electric motor operation possible.

ELECTROMAGNETIC FIELDS

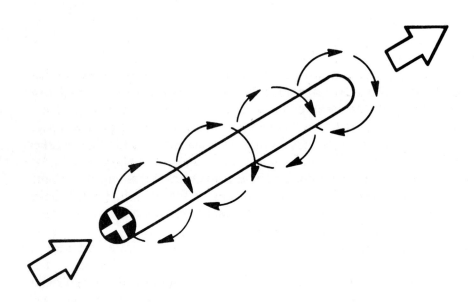

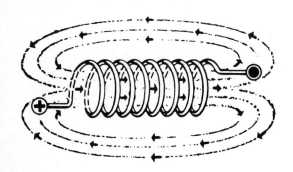 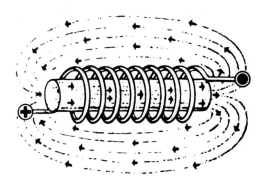

MAGNETIC FIELD SURROUNDING A CURRENT CARRYING CONDUCTOR

ELECTROMAGNETIC FIELDS

Electricity and magnetism are two separate but closely related forces. This is demonstrated by the fact that magnetic lines of force are produced around magnets, and also around conductors carrying electrical current. When electrical current is passed through a conductor, there will always be a magnetic field surrounding the conductor. The strength of this magnetic field depends upon the amount of current flow. The higher the amperage, the greater the magnetic strength.

If two conductors are arranged side by side the current passes through both conductors in the same direction, the magnetic field around each conductor will be in the same direction. As a result, the two magnetic fields will combine to form one stronger field surrounding both conductors. This causes the two conductors to be drawn together or attracted to each other. If the current is in opposite directions, the magnetic fields surrounding the two conductors will oppose each other and result in a repelling action. This is the principle involved in the operation of an electric motor, such as a starter motor on a vehicle.

If a conductor is wound into a coil, the current passing through it will flow in the same direction in all turns. The magnetic field produced by each turn combines with the field produced by adjacent turns, resulting in a strong continuous field lengthwise around and through the coil. The polarity of the field produced by the coil depends upon the direction of current flow and the direction in which the coil is wound. The strength of the magnetic field depends upon the number of wire loops and the amount of current passed through the coil. This is known as *ampere-turns*.

The strength of the magnetic field around the coil can be materially strengthened by placing a core of soft iron inside the coil. Because iron is a much better conductor for the magnetic lines of force than is air, the field becomes more concentrated and much stronger. Electromagnetic relays using this basic design are used in many applications in the electrical system of the automobile.

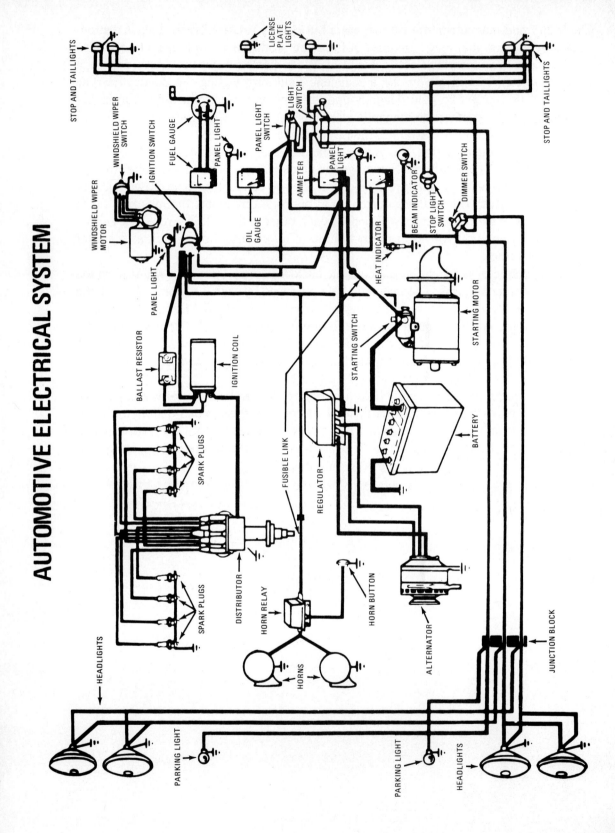

Figure No. 9 — Automotive Electrical System

AUTOMOTIVE ELECTRICAL SYSTEM

Figure No. 9 shows the charging system, starting system, the ignition system, the lighting system, and various accessory circuits. Today's modern automobile, with its high-compression engine and the addition of various accessories, has brought about an electrical system which has become increasingly complex. With the need for improved starting efficiency, greater demands on the ignition system, and larger generator capacity, new electrical units have been developed and others redesigned. Every unit of the automotive electrical system, whether it be a generator, a coil, a solenoid, or a voltage regulator, contains an electrical circuit and depends upon electricity in some form to do its work. The only way to determine the condition of these circuits is by accurately and efficiently testing their components.

Fusible Links and Wires

The function of the fusible link is to protect circuits that are not normally fused, such as charging circuit, ammeter, parking brake alarm, turn indicator circuit, cigarette lighter, and others.

One form of fusible link is inserted in the wiring harness, usually at a junction block in the engine compartment. Another form attaches the link to the insulated battery cable.

The fusible link is constructed of a fuse wire covered with a Hypalon insulation. The fuse wire is normally four wire gauges smaller than the circuit it is protecting. When the circuit is subjected to a short-circuit overload, the fusible link overheats and burns out, protecting instruments and wiring from damage. The Hypalon insulation, which is capable of withstanding extreme temperature, swells to about twice its normal size and assumes a "bubble" appearance. This indicates that the fuse link it contains has burned out.

Burned links are replaced by being cut out of the circuit and replaced with new links of the proper size after the cause of the trouble has been located and corrected.

Another form of this device is a fusible wire used on the charging-system voltage regulator on Chrysler Corporation vehicles. If the alternator field current draw should become excessive, as in the event of short circuiting during charging-systems tests or accidental grounding of the alternator field lead, the fusible wires will melt, thereby protecting the regulator, the alternator, and the wiring harness. These fusible wires are also replaceable after the cause of the overload is corrected.

Figure No. 10

AUTOMOTIVE GROUND CIRCUITS

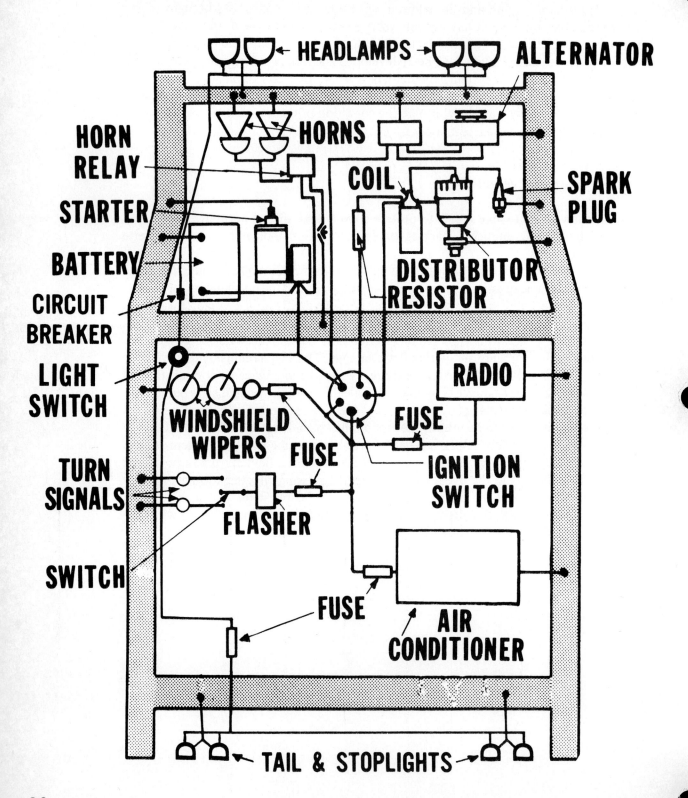

AUTOMOTIVE GROUND CIRCUITS

Every individual circuit in the automotive electrical system has both an insulated circuit and a ground circuit. The insulated circuit contains the source of electrical power.

The ground circuit, of equal importance to the insulated circuit, contains the ground connections and the metal parts of the vehicle that serve to complete the circuit. The metal parts of the vehicle are the frame, the firewall, the body and the ground straps.

The ground straps are an essential part of the ground circuit, since the engine and body are mounted on rubber mounts or biscuits. These rubber mounts absorb engine and road vibration and noise and prevent their being transmitted into the body. The ground straps are usually braided metal straps that are bolted between the engine and firewall and between the engine and the body or fender sheet metal.

All electrical system testing is divided into two groups—testing the insulated circuit and testing the ground circuit. Proper functioning of *both* circuits is essential to the efficient operation of every electrical unit. Examples of these test procedures will be covered in detail as the course progresses.

WIRING DIAGRAM

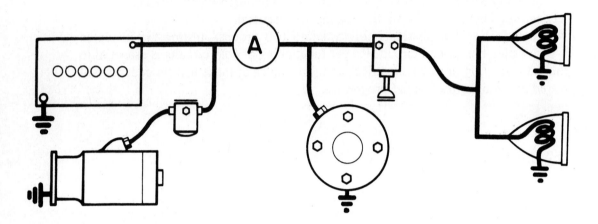

SCHEMATIC DIAGRAM

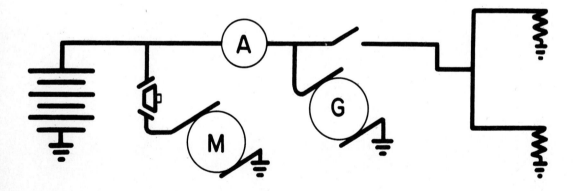

WIRING AND SCHEMATIC DIAGRAMS

To fully understand the illustrations that are found in the shop repair manuals, we should know the two basic types of diagrams. A wiring diagram is a pictorial view of an electrical circuit showing the components as they are actually connected.

A schematic diagram makes use of symbols to designate what the components are in the circuit. These symbols are used for clarification purposes and ease of understanding.

Figure No. 12

JUMPERS NEED A VARIETY OF TERMINALS

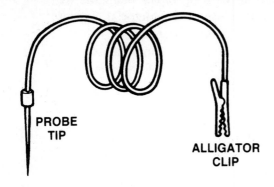

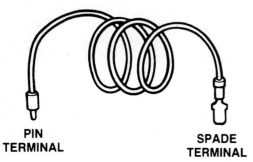

(Courtesy of Ford Motor Company.)

Figure No. 13

TYPICAL PROBE-TYPE TEST LIGHTS

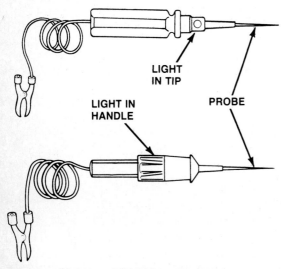

(Courtesy of Ford Motor Company.)

Figure No. 14

AMMETER IN USE

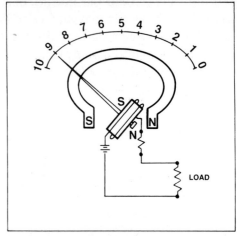

(Courtesy of Ford Motor Company.)

Figure No. 15

OHMMETER IN USE

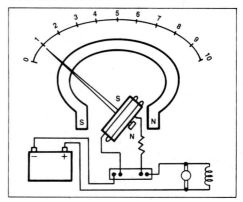

(Courtesy of Ford Motor Company.)

Figure No. 16

VOLTMETER IN USE

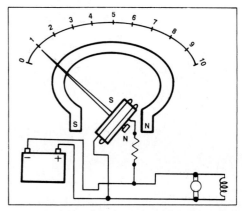

(Courtesy of Ford Motor Company.)

CIRCUIT TESTING TOOLS

The use of circuit testers is essential to the repair of any electrical system. The measurement of voltage, resistance (ohms), and current (amperes) allows the mechanic to determine if an electrical circuit is functioning properly.

Jumper Wire

The jumper wire (Figure No. 12) is the simplest device and can be made by attaching alligator clips to each end of a length wire. Because some tests require the use of more than one jumper, it is a good idea to construct several wires of various lengths and gauges.

The jumper wire is used to bridge an open circuit or bypass a suspected short in the circuit. Be careful *not* to connect the jumper between a load-carrying segment of a circuit and ground. Such a connection could cause an electrical fire and damage to the system being tested.

Test Light

A test light (Figure No. 13) is nothing more than a jumper wire with a 12-volt lamp spliced into its center. The light can be used to test for shorts or voltage in a circuit. By connecting between a live conductor and a good ground, the lamp will indicate the presence of voltage in the circuit. Remember, the test light is only an indicator and cannot be used to measure voltage.

Self-powered Test Light

Often referred to as a continuity tester, the self-powered test light consists of two test leads wired in series with a lamp and battery. This tester is used to check for grounds or continuity in a circuit. By connecting the self-powered tester leads between two points in a circuit, the bulb will glow indicating continuity.

The continuity tester must never be used on a live circuit, or damage to the system may result. Self-powered testers are not recommended for use with digital computer controls due to their sensitivity to power inputs.

Ammeter

By connecting an ammeter (Figure No. 14) in series with a circuit, the amount of current (amperes) flowing in the circuit can be measured. Ammeters will be internally or externally shunted in order to prevent exposure of the ammeter to full current flow.

Never connect an ammeter across a circuit; it must always be placed in series. Selecting the proper measurement scale is also critical. Failure to follow these precautions will damage the meter.

© Copyright 1986, Tune-Up Manufacturers Institute

Ohmmeter

The ohmmeter (Figure No. 15) is used to measure the resistance to flow in the circuit. Its operation is similar to a self-powered test light, except that the ohmmeter measures resistance in terms of ohms. The ohmmeter contains a battery that is used to send current through the circuit and it measures the drop across the conductor. Because of this, ohmmeters cannot be used when a conductor is live.

The ohmmeter is essential to the testing of variable resisters such as temperature and pressure sending units. Since senders are usually grounded through their threaded bodies, the ohmmeter can also be used as a continuity tester.

No check of an ignition system would be complete without a resistance test of the primary and secondary coil windings. It is little wonder that mechanics consider the ohmmeter to be one of the most valuable pieces in their tool box.

Voltmeter

The voltmeter (Figure No. 16) is used to measure the difference in electrical potential between two points in a circuit. Always connect this instrument in parallel with the circuit and make sure that the meter scale selected is capable of handling the maximum voltage potential of the circuit.

A battery voltage test best illustrates voltmeter usage. Select the lowest dc voltage scale that exceeds the normal 12-volt capacity of the battery. The 20-volt scale is most common. Connect the tester across the battery terminals (usually red to positive and black to negative). The voltmeter reading measures difference in electrical potential between the positive and negative sides of the battery.

Notes

Figure No. 17

Standard diode symbol.

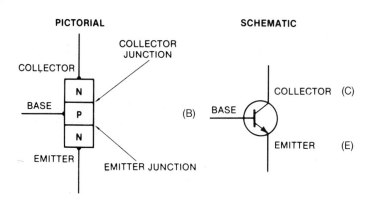

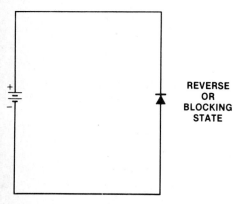

Diode connected in both its forward and reverse state.
(Courtesy Automotive Service Industry Association.)

Figure No. 18

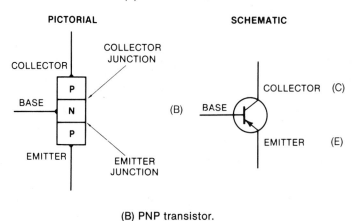

(A) NPN transistor.

(B) PNP transistor.

Standard transistor symbols and construction principles for both NPN and PNP types.
(Courtesy Automotive Service Industry Association.)

INTRODUCTION TO SOLID-STATE ELECTRONICS

Prior to the advent of solid-state electronics, the control of current flow in a system was accomplished by mechanical devices. Switches, solenoids, relays, and thermal resistors are capable components, but due to the nature of mechanical movement and resistor heating, their life span and accuracy are limited. Solid-state components are devices that control the flow of current without relying on any mechanical moving parts or heat-generating resistors.

The invention of semiconductors made solid-state electronics possible. While semiconductor technology became available in 1948, its use on the automobile began in the late 1950s. The delay in vehicular application of the semiconductor was due to the harsh environment of the automobile. Years of development and testing were required to make solid-state devices compatible with this environment. Each year the use of semiconductors has expanded, and it will continue to expand as more mechanical components are replaced by electronics. The two most common semiconductors used in solid-state devices are the diode and transistor.

Diodes

The simplest way to understand a diode is to look at it as a one-way check valve. Diodes are used to allow normal current flow in one direction, but limit or block current flow in the opposite direction (Figure No. 17).

The diode is constructed of two different semiconductor properties. One wafer is composed of negative (N) materials, while the other is a positive (P) material. By connecting the positive terminal (anode) of a battery to the P material and the negative terminal (cathode) to the N material, normal current flow is permitted. This arrangement is referred to as a *forward-biased* connection. If the anode and cathode connections are switched, a *reverse bias* is created and current flow is blocked or greatly restricted. This behavior allows the diode to be used in the conversion of ac to dc current (rectifier) and for the protection of components from voltage spikes in a system.

The most accurate method of testing a diode involves the use of an ohmmeter. Switch the meter to the lowest scale and read the resistance in both directions (reversing the leads). A properly functioning diode will indicate very low resistance in one direction and extremely high resistance in the opposite direction. Should both readings show continuity (zero), the diode is shorted, and if both readings indicate infinity, the diode is open.

A self-powered test light can also be used. The light should glow in one direction and remain out in the other. If it lights in both directions, it is shorted, and if it remains out in both directions, the diode is open.

It should also be noted that a diode will occasionally check good when cold and fail to perform when hot. Some test procedures call for the diode to be warmed with a heat gun prior to testing. This procedure is often called for when repairing computerized engine control systems. Consult the repair manual.

Zener Diodes

Certain circuit designs call for diode construction that allows current conduction in both directions. Such a diode is known as a zener diode. Forward-biased operation is the same as a normal diode, but reverse bias can be calibrated to allow reverse current flow when a certain voltage level is reached. For example, a particular zener diode may block reverse flow when the voltage is below 10 volts, but become conductive when the reverse-biased voltage reaches 10 or greater volts. Zener diodes are generally used in control circuits for the switching of transistors.

Transistors

A transistor is a semiconductor device used to control the flow of current in a circuit. It is similar to a relay, in that a low-current signal is used to switch a higher-current application. Since a transistor has no mechanical moving parts, it is more reliable than a solenoid-activated relay.

Like a diode, the transistor is constructed of negative (*N*-type) and positive (*P*-type) materials (Figure No. 18). The arrangement of these materials into alternate sections will determine the polarity of the transistor. The materials can be arranged to create a *PNP* or an *NPN* transistor. The center section serves as the base, which controls the flow of current through the transistor. The type of material used to form the base section determines the polarity.

The transistor symbol is illustrated in Figure No. 18. The circle represents the container. The letters represent the three sectional materials of which the transistor is composed: E is the emitter, C is the collector, and B is the base. The arrow on the emitter symbol indicates the direction of current flow through the transistor.

There are two important factors relative to the manner in which a transistor works. First, the emitter–collector circuit is the main current-carrying circuit. Second, the emitter–collector circuit becomes conductive only when there is current flowing through the emitter–base circuit. Although the current flow in the base circuit may be only a fraction of the current flow in the collector circuit, the collector circuit cannot exist without the base circuit. Therefore, the interruption of the low current flow through the base circuit will cause a stoppage in the high current flow in the collector circuit. In this manner, the base circuit triggers the transistor, turning it on or off.

Notes

Notes

3

BATTERY

BATTERY VISUAL CHECKS

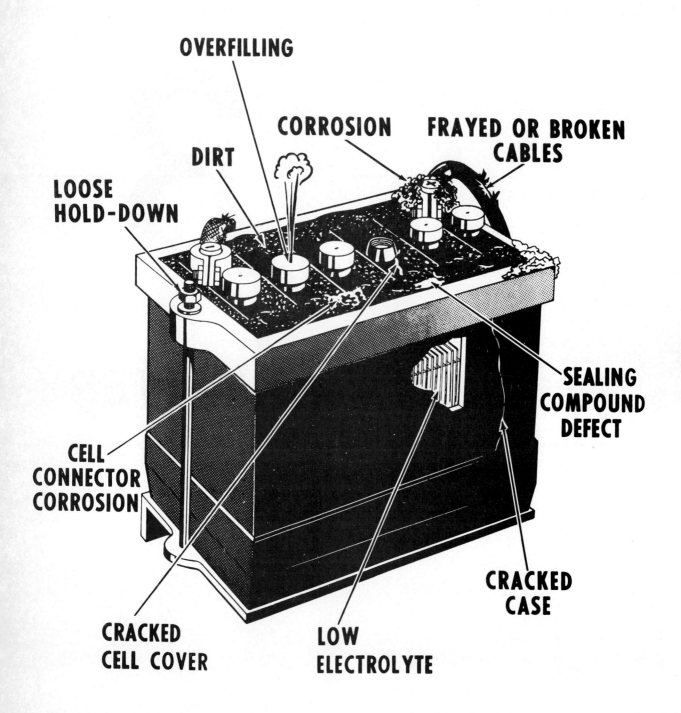

Figure No. 19

BATTERY VISUAL CHECKS

It has often been stated that the man who tests the most batteries, sells the most batteries.

Battery care and testing are relatively simple. A basic knowledge of how a battery is constructed, how it works, along with a few pieces of test equipment and simple test procedures, will provide any serviceman with the essentials he needs to provide excellent service in this money-making phase of tune-up.

The primary function of the battery is to provide power to operate the starting motor. It must also supply the ignition current during the starting period and accomplish this even under adverse conditions of temperature and other factors.

The battery can also serve, for a limited time, as a source of current to satisfy the electrical demands of the vehicle which are in excess of the output of the generator.

Batteries used in automobiles are known as *storage* batteries. This term storage battery is sometimes misinterpreted. A battery does not store electricity, but does store energy in a chemical form. It accomplishes its task by a chemical process which takes place inside a battery when it is connected to a complete circuit.

Basically stated, a battery is composed of two dissimilar metals in the presence of an acid. The battery is constructed of a series of positive and negative plates. These plates are insulated from each other by means of separators. All the positive plates are interconnected, and all the negative plates are interconnected. These interconnected series of positive and negative plates are submerged in the container filled with a sulfuric acid and water solution known as *electrolyte*.

The first test of a battery is a visual inspection. If a battery is cracked or otherwise defective, it must be discarded. If the electrolyte level in the battery is low, or if the ground connections or insulated connections are defective, the battery cannot operate efficiently. It is also very important that the battery be kept clean. Dirt and moisture can serve as a conductor and slowly discharge the battery over a period of time.

When activating a dry-charge battery, follow the battery manufacturer's service procedure and be sure to fill each cell properly with the electrolyte supplied. Apply a warm-up charge of 15 amperes for 10 minutes after activating the battery, when so instructed. Observe charging cautions.

Dry-charge batteries will be damaged if moisture enters the battery while it is stored. Always store dry-charge batteries in the coolest and driest location possible.

Figure No. 20

BATTERY LEAKAGE TEST

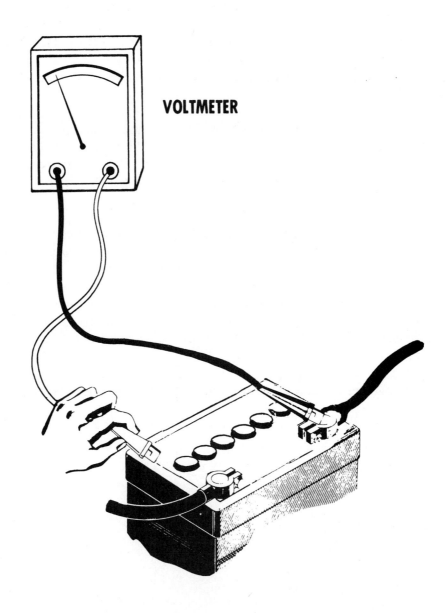

BATTERY LEAKAGE TEST

By using a voltmeter, a tune-up specialist can show his customer that a battery with a dirty top is actually leaking current and may become self-discharged.

A leakage test is performed by clipping the negative voltmeter lead to the battery negative terminal. The positive voltmeter lead should be moved over the insulated surface of the top of the battery. Any reading is an indication of electrical leakage. This undesirable electrical path is composed largely of electrolyte which has been expelled from the battery through the fill cap vents by the charging action of the generator. Electrolyte is, of course, a conductor of electricity.

If meter readings indicate any electrical leakage, the battery top should be washed with a solution composed of one spoonful of baking soda mixed in a pint of water. After the bubbling action induced by acid neutralization stops, rinse the battery top with clean water and dry the battery.

Corrosion accumulation around the cable clamps should be removed with a brush and they should be washed with the same solution.

Figure No. 21

BATTERY SPECIFIC GRAVITY

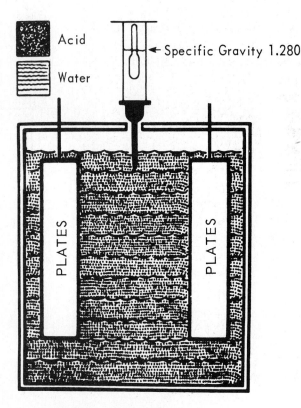

FULLY CHARGED
Acid in water gives electrolyte specific gravity of 1.280

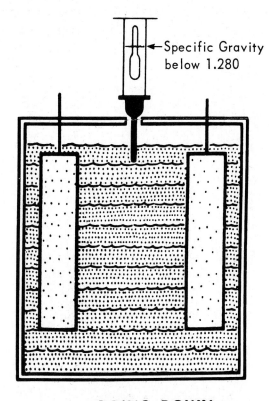

GOING DOWN
As battery discharges, acid begins to lodge in plates. Specific gravity drops.

© Copyright 1986, Tune-Up Manufacturers Institute

BATTERY SPECIFIC GRAVITY TEST

By using a hydrometer, the specific gravity of the electrolyte solution in a battery can be determined. The battery specific gravity is an indication of the battery state of charge. If the state of charge is low, the hydrometer will read low. If the state of charge is high, the hydrometer will read high. As an example, a reading from 1.260 to 1.280 indicates a fully charged battery. A reading from 1.200 to 1.220 indicates a battery that is in a half-charged condition. Readings below 1.200 indicate that a battery is in a discharged condition and cannot give satisfactory service.

The definition of specific gravity is the weight of a liquid compared to the weight of an equal volume of water. The specific gravity of chemically pure water at 80°F is 1. Therefore, by knowing the specific gravity of sulfuric acid, we can accurately measure the ratio of sulfuric acid to water in the battery electrolyte solution.

When a battery is in a full state of charge, the negative plates are basically sponge lead, the positive plates are lead peroxide, and the electrolyte has a maximum acid content and a minimum water content.

As the battery is discharging, the chemical action taking place reduces the acid content in the electrolyte and increases the water content, while both negative and positive plates are gradually changing to lead sulfate.

When the battery is in a state of discharge, the electrolyte is very weak, since it now has minimum acid content and a maximum water content, and both plates are predominately lead sulfate. The battery now ceases to function because the plates are now basically two similar metals in the presence of water, instead of two dissimilar metals in the presence of an acid.

During the charging process, the chemical action that occurred during the battery discharge is reversed. The lead sulfate on the plates is gradually decomposed, changing the negative plates back to sponge lead and the positive plates back to lead peroxide. The acid is redeposited in the electrolyte, returning it to full strength. The battery is now again capable of performing all its functions.

After activating a dry-charge battery, check the specific gravity. The gravity reading should be 1.260 or slightly higher. If the electrolyte level drops shortly after the initial fill, due to the plates and separators absorbing some of the solution, add more electrolyte to bring the solution up to the proper level. When so instructed, charge the battery at 15 amperes for 10 minutes before installing the battery to assure a full charge.

No. 22

TEMPERATURE CORRECTED HYDROMETER

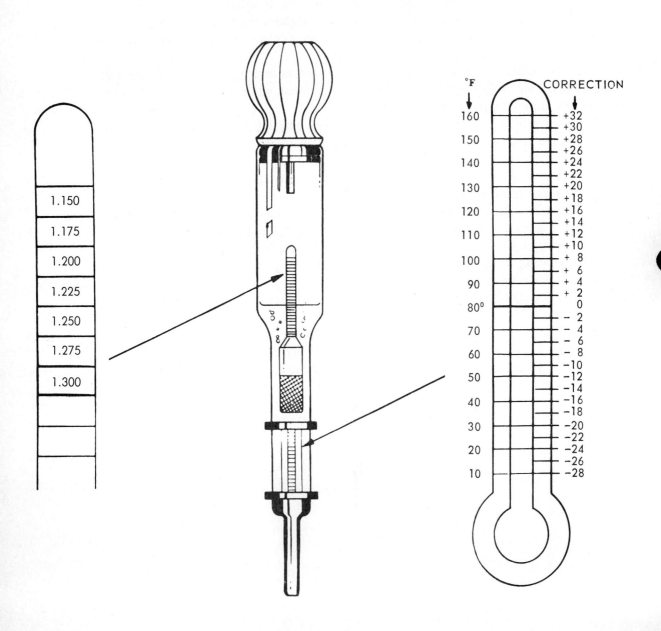

TEMPERATURE-CORRECTED HYDROMETER

Hydrometer floats are calibrated to indicate correctly only at 80°F temperature. If used at any other temperature, a correction factor must be applied. The reason for this lies in the fact that a liquid expands when it is heated and shrinks when it is cooled. This will cause a change in the density of the electrolyte solution, which will raise or lower the specific gravity reading.

A thermometer is built into the temperature-compensating-type hydrometer. The scale of this thermometer indicates the temperature of the solution. This reading should be used so that the proper temperature correction factor can be applied.

The table is based on an electrolyte temperature of 80°F. For other temperatures, correct the indicated reading by adding 4 points (0.004) for each 10° above 80°F, and subtracting 4 points for each 10° that the electrolyte temperature is below 80°F.

For example, a specific gravity reading of 1.230 is obtained at a solution temperature of 10°F. If the electrolyte temperature is disregarded, the reading of 1.230 may be considered as low but acceptable. When the reading is temperature corrected, the true reading of 1.202 (7 × 4 = 28 from 1.230) reveals that the battery is actually very low and definitely in need of charging.

A specific gravity reading of 1.235 is obtained at a solution temperature of 120°F. The reading itself may be interpreted as being rather low, but when temperature corrected the reading is actually 1.251 (4 × 4 = 16 added to 1.235). This specific gravity may be high enough for the battery to be restored to full charge by the car's generator.

These examples indicate the importance of temperature correcting specific gravity readings to accurately interpret the true state of battery charge.

To accurately test the true condition of the battery, a light-load test or a capacity test should be conducted after the specific gravity has been tested.

Figure No. 23

BATTERY CAPACITY (LOAD) TEST

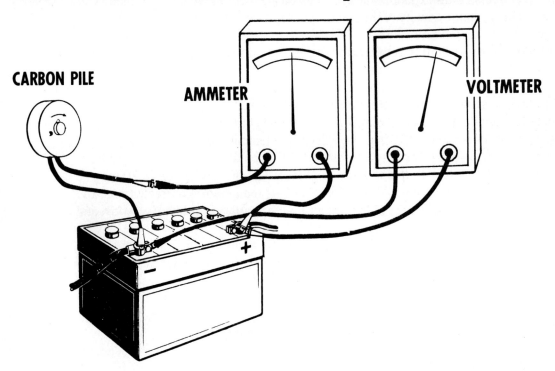

THREE MINUTE CHARGE TEST

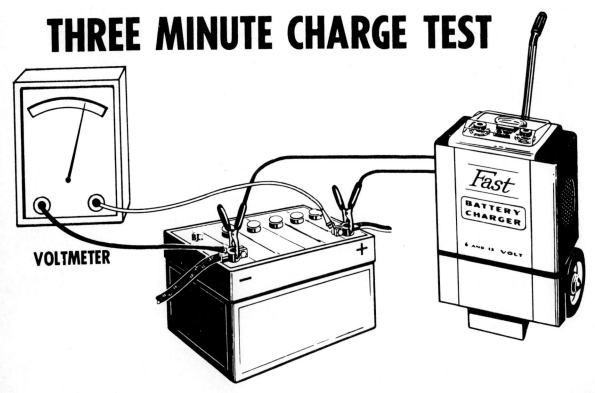

BATTERY CAPACITY (LOAD) AND THREE-MINUTE CHARGE TESTS

Most engine starting failures are caused by the inability of a battery to maintain a voltage high enough to provide effective ignition while cranking a cold engine.

Battery Capacity Test

The function of the battery capacity test is to duplicate the battery drain of a cold engine start while observing the battery's ability to maintain voltage. A battery that passes the capacity test will provide dependable performance.

The *battery/starter tester* has an ammeter, a voltmeter, and a carbon pile, which is a battery loading device. The charged battery is discharged at a rate of three times its ampere-hour rating for 15 seconds while its voltage is observed. The voltage of a 12-volt battery should not drop below 9.0 volts, or that of a 6-volt battery below 4.5 volts. A reading below this specification indicates a defective battery that should be replaced.

Three Minute Charge Test

A battery that is less than fully charged may be tested with a fast battery charger and a *3 minute charge test*. A fast battery charger is used in conjunction with the battery/starter tester for this test. Fast charge the battery three minutes at not more than 40 amps for a 12-volt battery, and 75 amps for a 6-volt battery. With the charger in operation, observe the voltage of the battery. If the voltage exceeds 15.5 volts for a 12-volt battery, or 7.75 volts for a 6-volt battery, the battery is sulfated or worn out. This indicates that the plates will no longer accept a charge under normal conditions, and the battery should be discarded.

Be sure to observe all precautions relative to working around a battery while it is being charged. Explosive hydrogen gas is liberated from the electrolyte while the battery is being charged. Sparks from a lighted cigarette or from charger clamps being disconnected while current is still flowing may cause an explosion that will destroy the battery and possibly inflict personal injury.

Also *be sure* that all the precautions that are relative to working on a battery installed in a car that is equipped with an alternator are observed. These precautions will be fully covered in our discussion on alternator charging systems.

A maintenance-free battery with a one-piece, hard-top cover must be tested with the capacity test, since individual cell tests cannot be conducted. Under no circumstance should there be an attempt to pierce the one-piece battery cover with meter prods to conduct an individual cell test.

Maintenance-Free Batteries

There are two types of maintenance-free batteries, maintenance-free batteries and low-maintenance batteries. The major manufacturers that currently make maintenance-free batteries with *no antimony* are Delco-Remy, Gould, Prestolite, and General. Those making batteries using *low antimony* call their batteries Low Maintenance. The major manufacturers of these are currently Globe-Union, ESB Brands, Inc. (Exide and Willard), Prestolite, and General. Any type of maintenance-free battery that is sealed cannot be tested using a 3-minute charge test.

Important Testing Data

There have been instances where maintenance-free batteries have been judged to be defective based on the fast-charge sulfation test. This test should be valid provided it is properly used. The following rules apply to all batteries, not just maintenance-free batteries.

1. Any automotive battery can become sulfated if it is left in a discharged or partially charged state for any length of time. Sulfation is caused as follows: When a battery is being discharged, the sulfuric acid enters the plates. The electrolyte becomes weaker and weaker as discharge proceeds, and more and more acid enters the plates. If this acid is not all moved back from the plates to the electrolyte by fully charging the battery, the acid left in the plates will eventually harden (turn to sulfate) and can never be recovered. The on-charge test determines if this has happened.

2. The fast-charge test should never be performed on any battery unless it has failed the load test, for a charged battery acts the same as a sulfated battery when a fast charge is put into it. The voltage increases rapidly as high as 18 to 20 volts.

3. When performing the 3-minute fast-charge test, the charge rate should be set as near to 40 amps as possible, but not over 40 amps. With the fast charger, this can easily be done. On some chargers, it cannot be done. The charge rate should never be increased after the initial setting is made.

If these rules are followed, the sulfation test is valid. Remember, until further notice, Delco-Remy maintenance-free batteries use 16.5 volts as the upper limit. All others use 15.5 volts.

Maintenance-Free Battery Test Procedure

Suggested test procedure for maintenance-free batteries, other than Delco-Remy and Gould, is as follows:

1. Stabilize batteries at room temperature.
2. Remove the surface charge by applying a load of 300 amperes for 15 seconds.
3. Apply a load equal to one-half cold cranking current at 0°F for 15 seconds.

Test Indications

1. If voltage after 15 seconds with load applied is 9.6 or more, battery is serviceable.
2. If voltage is below 9.6 volts, fully charge battery and retest, using steps 1, 2, and 3.
3. If after recharging and load testing load voltage is still below 9.6, battery should be replaced.

Note: The 3-minute sulfation test can be made if desired on those batteries that will not accept a charge in the vehicle with a good charging system. Be sure to follow the proper procedure. The 3-minute test procedure will eventually be revised to cover maintenance-free batteries. At this time it applies only to conventional batteries.

Battery Rating Methods

Over the years, many methods have been devised to specify the capacity or electrical size of batteries. Presently, only three methods are commonly used: (1) the amp-hour method, (2) cold cranking performance, and (3) reserve capacity. In addition to electrical rating methods, batteries are also arranged according to their physical size by *group numbers*. Batteries with the same group number have the same dimensions and are physically interchangeable. However, they may have widely varying electrical capacities and for this reason are not always interchangeable.

Amp-Hour Method

The amp-hour method has been used for many years, although it is being gradually replaced by the cold cranking and reserve capacity ratings. A battery's amp-hour rating is determined by discharging a fresh, fully charged battery at a constant rate so selected that at the end of 20 hours the voltage will have fallen to 1.75 volts per cell (or 10.5 volts for a 12-volt battery). This discharge current, times 20 hours, gives the battery's amp-hour rating. For example, if the required discharge current was 3.0 amperes, the battery would be rated at 3×20 or 60 amp-hours. It should be noted that this does *not* mean that such a battery can be discharged at 60 amperes for 1 hour, or any other combination, with the same results. When replacing batteries, always replace with the specified, or higher, amp-hour battery.

Cold Cranking Performance

A more recent rating method is designed to show a battery's cold weather cranking ability. A cold cranking rating shows how many *amperes* can be drawn from a battery at 0°F for 30 seconds before its voltage drops below 1.2 volts per cell (or 7.2 volts for a 12-volt battery). As a rough rule of thumb, a battery's cold cranking rating in amperes should approximate the engine's displacement in cubic inches. Most new batteries have this rating, and sometimes the amp-hour rating, imprinted on the battery cover. There is no convenient way to convert between amp-hour ratings and cold cranking ratings.

Reserve Capacity

The reserve capacity rating is specified in *minutes* and indicates for how long a vehicle can be driven with battery only in the event of charging system failure. The rating is established by noting how long it takes a fully charged battery (at 80°F) to drop to 1.75 volts per cell (or 10.5 volts for a 12-volt battery) at a constant 25-ampere discharge rate. This discharge rate represents a typical nighttime electrical load with headlights and heater. Thus a battery with an 80-minute reserve capacity rating could keep a vehicle with a defective charging system running for a least one hour and 20 minutes. Operating time without the headlights and heater, of course, would be even longer. It must be remembered, however, that this rating assumes the battery to be fully charged initially. A partly charged battery may not provide the full specified operating time.

4

STARTING SYSTEM

Figure No. 24

STARTING CIRCUIT

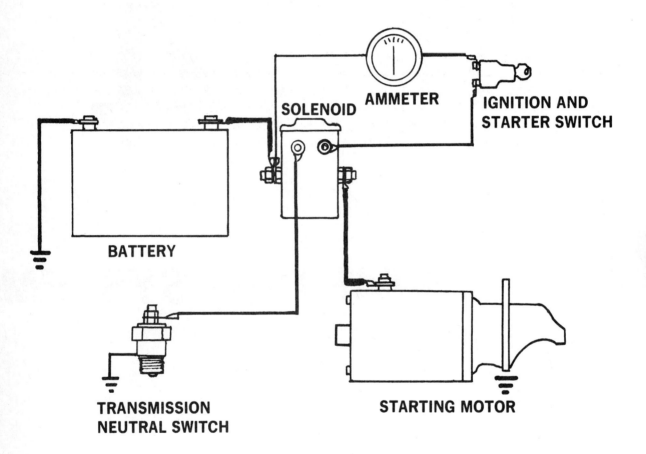

STARTING CIRCUIT

A starting motor, starting switch, battery, and cables, which comprise the starting circuit, provide the power for cranking the engine. The motor receives electrical power from the battery and converts it into mechanical power, which is transmitted to the engine through a drive pinion gear and the flywheel ring gear. The starting switch controls the operation by making and breaking the circuit between the battery and the motor.

The battery, starting motor, starting switch, and the wiring are all designed for the high current flow needed to produce efficient cranking power. The condition of these components is extremely critical, as even a small amount of resistance can cause a marked reduction in cranking ability. This indicates the necessity for testing the various components in the starting system.

The starter is connected in series with the battery. This is known as the *high-amperage circuit*. The solenoid is the switch between the battery and the starter, and is activated by a low-current-carrying circuit known as the *control circuit*.

When using a solenoid starter switch (remote-control switch), *follow the hook-up instructions* that came with the switch to avoid the possibility of damaging the starter control circuit on the vehicle. For example, if the remote-control switch lead is clipped to the solenoid terminal of the transmission neutral switch, the neutral switch will be burned out when the remote-control switch is actuated. Also observe the precautions for remote-control switch use that are covered in the compression test of the tune-up procedure in Chapter 12.

Figure No. 25

STARTING MOTOR

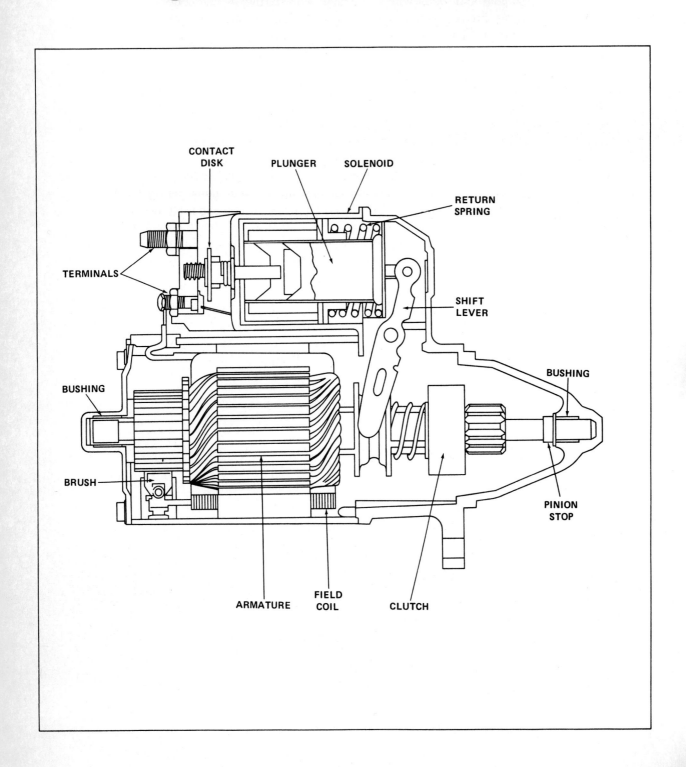

THE STARTING MOTOR

The starting motor converts electrical energy from the battery into mechanical power and transmits the power to the engine. It consists of a frame and field, armature, commutator end head, drive end head or housing, and a steel frame which supports the components of the motor and also forms part of the magnetic path. Field windings, consisting of coils of copper strips or wires, produce a magnetic field which is conducted through steel pole shoes to the armature. The armature core is also steel to complete the magnetic circuit. Copper conductors are installed in lengthwise slots around the armature core and are connected to a commutator consisting of a number of copper segments insulated from each other and from the armature shaft. The armature turns in bearings mounted at each end of the motor, and brushes in the commutator end make electrical contact with the revolving commutator.

Starting motors vary as to type of connections of the internal wiring and in the method of transmitting the power to the flywheel. Four poles and brushes are commonly used for 6-and 12-volt equipment. Usually, the motor is mounted by a flange on the pinion housing or drive end head. Motors designed for inboard meshing, or which have the drive mechanism entirely within the flywheel housing, use a head to close the drive end of the frame and support the armature. In some motors, a bearing is installed near the middle of the armature to provide additional support.

STARTING MOTOR DRIVES

Figure No. 26

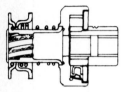

Figure No. 27

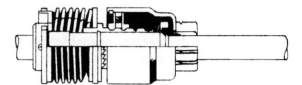

Figure No. 28

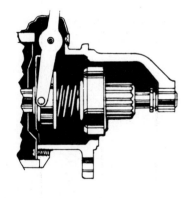

OVERRUNNING CLUTCH

Figure No. 29

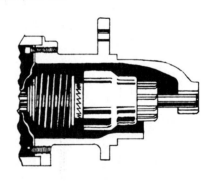

INERTIAL ENGAGEMENT

STARTING MOTOR DRIVES

To transmit the power to the flywheel of the engine, starting motors use either an overrunning clutch or an inertial engagement drive. Both of the drives serve two purposes. The first is to engage and disengage the starting motor and the engine flywheel. The second purpose of the drive is to provide a gear reduction between the starting motor and the flywheel so that the motor will have enough power to turn the engine fast enough for it to start.

Overrunning-Clutch-Type Drive

The overrunning clutch drive is used extensively and is shifted to and from the meshed position, either by a solenoid or manually.

The overrunning clutch transmits cranking torque from the starting motor armature to the flywheel when the engine is being cranked and disconnects the armature from the pinion gear after the engine starts. This is accomplished by an arrangement of cams and rollers within the clutch assembly. Under cranking conditions, each roller is forced into the narrow space between the cam and the pinion collar. This locks the entire assembly so that the power is transmitted from the armature to the flywheel.

After the engine starts, the flywheel tends to rotate faster than the motor. This action causes the rollers to move back to the released position, and permits the pinion to rotate freely with respect to the shell. During the shift to the cranking position, if the pinion teeth should butt against the ring gear teeth, the shifting lever compresses the meshing spring, permitting the shift lever to move to full operating position. This full operating position closes the starting motor switch contacts. Immediately, the armature and pinion rotate, unlocking the butting teeth, and the pinion enters the ring gear to complete engine cranking.

Inertial Engagement

The inertial engagement drive has been built in a number of different forms, but all operate by a combination of screw action and inertia. The pinion gear assembly is mounted on a threaded sleeve, which is driven by the armature, either by splines within the drive sleeve or by a spring which has one end connected to the armature shaft and the other connected to the pinion shaft. When the starting motor circuit is closed, the armature revolves, turning the drive sleeve within the pinion. The screw action forces the gear into mesh with the flywheel gear. The sudden shock of the meshing is absorbed by the assembly spring. When the engine starts, the pinion is driven faster than the shaft and is then threaded back along the screw shaft out of mesh with the flywheel. Clutch-type inertial engagement drives operate on the same principle, but they are designed for heavy-duty applications and incorporate a friction clutch which slips when overloads occur.

Figure No. 30

SOLENOID STARTING SWITCH

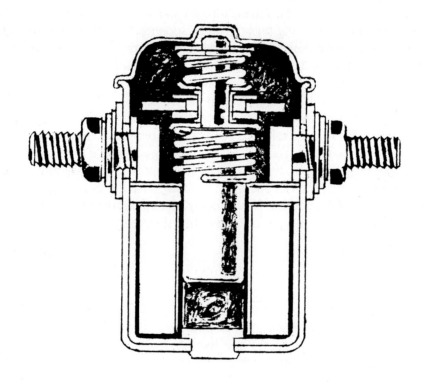

STARTING SWITCH

The starting switch controls the starting motor operation by making and breaking the circuit between the storage battery and motor. It is designed to carry the high current required by the starting motor with minimum voltage loss due to contact resistance, and also to make or break the circuit with positive action to reduce arcing and possible switch or motor failure.

Most passenger cars use some form of solenoid switch. Solenoid starting switches close the starting circuit through the magnetic pull of a solenoid on a steel core which carries the contact disc. The contact disc strikes two contacts that are connected to the external circuit. The contact disc on solenoid switches is mounted through a spring to make the opening and closing more positive and to align the disc with the contacts to give full surface contact. The switch illustrated in Figure No. 30 is constructed in this manner and has only one purpose, that of making and breaking the starting circuit. It is used with motors having a Bendix drive and cannot be used with an overrunning clutch.

Some solenoid switches have their winding circuit completed to ground internally. Others are grounded through the generator armature or a transmission neutral switch. These last two methods are used on many cars as a safety feature.

Figure No. 31

STARTING MOTOR AMPERAGE DRAW TEST

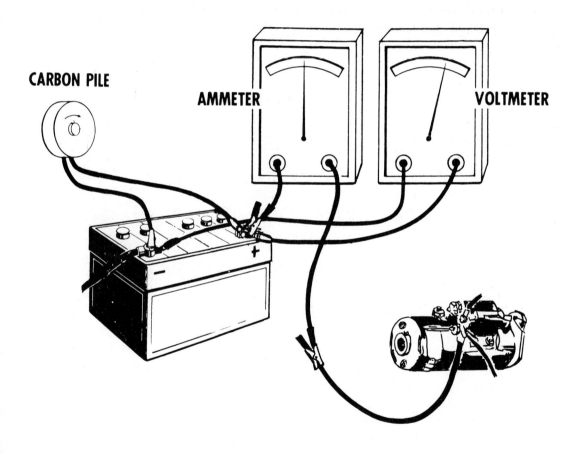

STARTING MOTOR AMPERAGE DRAW TEST

The starting motor amperage draw test is an on-the-vehicle check conducted with a battery starter tester to detect trouble in the starting motor. If the starting motor is operating properly, its amperage draw will be within specifications, 150 to 200 amperes, for example, and the cranking speed will also be normal, for example, 180 to 250 rpm.

So that this test can be considered a true test of the starting motor, other influencing variables will be considered to be within limits. The battery is more than three-quarters charged; the cable clamps are clean and tight; the engine oil is of the recommended viscosity; and the engine temperature is normalized. Further, it is considered that the starter solenoid connections are tight, the ignition switch is functioning properly, the starter is insulated, and ground circuits are normal.

The *battery starter tester* is designed with a high-reading ammeter, a voltmeter, and a carbon pile, which is a high-capacity variable resistor.

The starting motor amperage draw test is conducted as follows:

1. Connect the tester leads.
2. Connect a jumper lead from the distributor primary terminal to ground to prevent the engine from starting. Some manufacturers recommend disconnecting the primary lead from the coil; others suggest grounding the coil secondary lead.
3. Crank the engine and accurately observe the *exact* voltage indicated on the voltmeter.
4. Without cranking the engine, turn the tester carbon pile control until the voltmeter again reads the *exact* voltage it did while the engine was being cranked. Then read the ampere flow on the ammeter and release the tester carbon pile control. The ammeter reading is the starting motor amperage draw and should be within the manufacturer's specifications.

Higher-than-normal current draw, usually associated with slow cranking speed, is an indication of mechanical or electrical trouble in the starting motor. Remove the starter for repair or replacement.

Lower-than-normal current flow, also associated with slow cranking speed, indicates high resistance. This condition is caused by poor connections in the field or armature circuits, poor brush and commutator contact, or a defective commutator condition. Starting motor removal for service or replacement will be required.

As a word of caution—do not be too quick to condemn the starting motor unless you are certain the other factors, previously mentioned, are known to be functioning properly.

© Copyright 1986, Tune-Up Manufacturers Institute

Figure No. 32

STARTER INSULATED CIRCUIT TEST

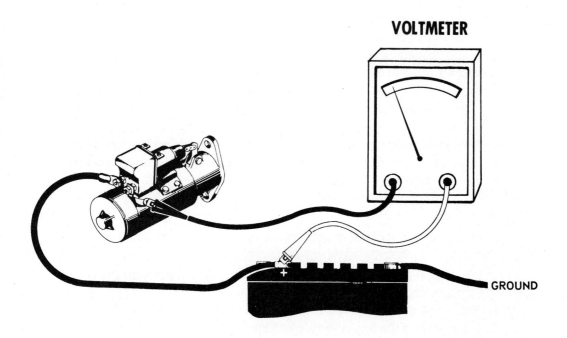

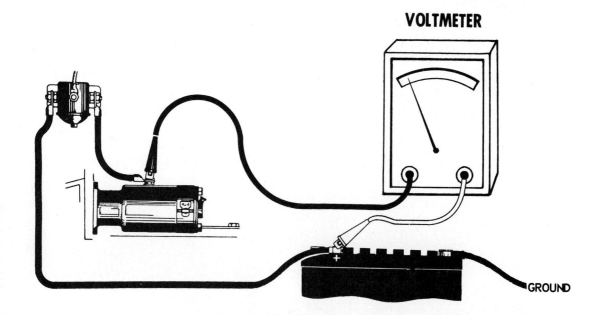

STARTER INSULATED CIRCUIT TEST

Loose or dirty connections or defective cables represent a power loss between the battery and the starter. Circuit resistance tests are made to determine if the insulated cable, switches, and ground connections can carry the current demanded by the starter. This resistance is indicated by a voltage drop. *Voltage drop* is the voltage expended in overcoming resistance in a given circuit. Permissible voltage drop in the average starter insulated circuit is 0.3 volt. The voltage drop allowed in this circuit is 0.1 volt per cable or switch. If the resistance is excessive, the result will be an extreme power loss in the starting system. When a starter motor is in operation, the high-amperage draw magnifies this seemingly low resistance value. This greatly reduces the efficiency of the entire starting system. Therefore, the circuit resistance tests are performed while the system is under normal cranking load.

To test the insulated circuit, one voltmeter lead is connected to the positive battery terminal, and the other voltmeter lead is connected to the large armature terminal on the starting motor or solenoid. Crank the engine and observe the voltmeter. If the voltage drop for the entire circuit is not excessive, objectionable resistance does not exist, and no further testing is necessary. If the test results indicate excessive resistance, separate detailed tests of each component in the circuit must be conducted.

Figure No. 33

STARTER GROUND CIRCUIT TEST

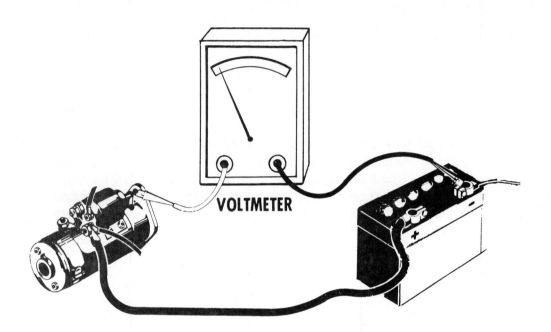

SOLENOID SWITCH CIRCUIT RESISTANCE TEST

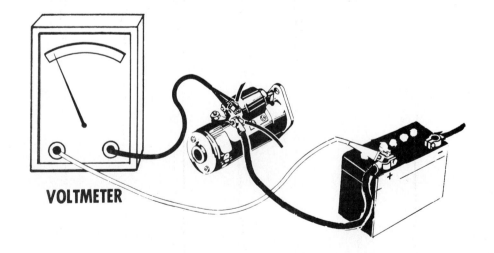

STARTER GROUND CIRCUIT TEST

To test the starter ground circuit, connect one voltmeter lead to the ground post of the battery and the other lead to a good ground on the starter motor. Operate the starter motor and note the voltage drop. Permissible voltage drop is 0.2 volt.

Voltage Drop

To make the solenoid circuit resistance test, connect one voltmeter lead to the positive battery post and the other lead to the proper control circuit terminal on the solenoid. Again note the voltage drop with starting system under load. Permissible voltage drop is 0.2 volt.

A higher-than-permissible voltage drop reading indicates at least one point of high resistance in the circuit. Separate detailed tests of each component and each connection in the circuit must be conducted. When the point of high resistance is located, the poor condition must be cleaned or tightened, or the defective unit or cable must be replaced, as the case may be.

Notes

5

CHARGING SYSTEM

CONSTANT VOLTAGE CHARGING SYSTEM

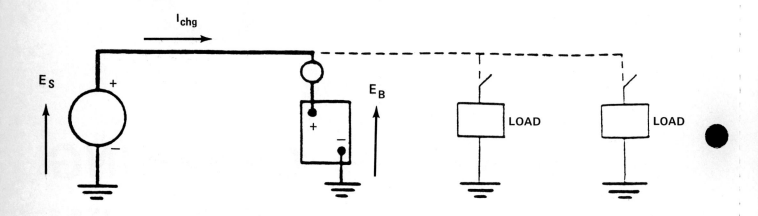

CONSTANT VOLTAGE CHARGING SYSTEM

The constant voltage charging system has been in use on vehicles for approximately 50 years and operates on the fact that the terminal voltage of a battery rises as it is charged. A battery at rest will have a voltage of 12.6 volts. When a battery is being charged by having a current pushed internally in the battery from positive to negative, its voltage will rise to approximately 14.4 volts when the battery is fully charged. With a charging system operating at 14.4 volts, the battery accepts a charge current until the battery voltage reaches approximately 14.3 volts. As battery and charging system voltage reach a common level, the ability of the battery to take a charge diminishes. At this point, the battery will only accept a small "trickle" charge.

To illustrate the operation of this type system, look at the system operation in three states of battery charge and voltage. The charging current can be determined by the following formula:

$$I_{CHG} = \frac{E_S - E_B}{R_{CS}}$$

where I_{CHG} = the battery charging current
E_S = source voltage = 14.2 volts
E_B = battery voltage
R_{CS} = total circuit resistance = 0.05 ohms

When the engine is first started, the battery voltage is 12.6 volts and the charging current is

$$I = \frac{14.2 - 12.6}{0.05} = 32 \text{ amperes (A)}$$

This charging current causes the battery voltage to rise, and when it reaches 13.4 volts the charging current will be

$$I = \frac{14.2 - 13.4}{0.05} = 16 \text{ A}$$

When the battery voltage reaches 14.1 volts, the charging current will be

$$I = \frac{14.2 - 14.1}{0.05} = 2 \text{ A}$$

This action is readily observed on vehicles equipped with an ammeter. After starting the vehicle, a high charge is observed, and in time it will decrease to a very low value. The length of time it takes the charge to reduce to a very low value will depend on the state of the battery charge. If the battery is badly discharged, the high charge will remain for a considerable period of time before reducing, whereas if the battery is in a high state of charge, the charging rate will taper off to the trickle charge in a very short period of time.

The system works very well provided a proper voltage can be maintained by the charging source, and the primary purpose of troubleshooting the system is to determine if the source is capable of developing and maintaining a proper system voltage.

© Copyright 1986, Tune-Up Manufacturers Institute

Figure No. 35

ALTERNATOR SCHEMATIC

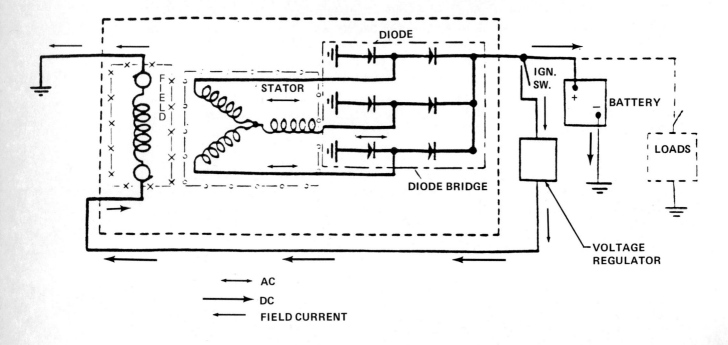

← → AC
→ DC
← FIELD CURRENT

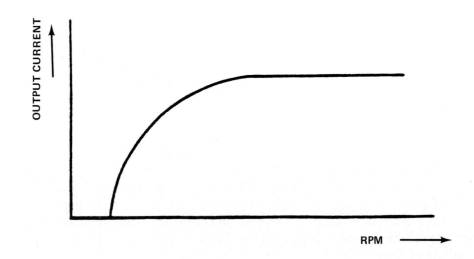

ALTERNATOR OPERATING PRINCIPLES

All rotating electrical charging sources create ac. With the advent of low-cost, highly reliable, efficient silicon diodes, the alternator became the charging source on vehicle systems.

The voltage that creates the charging current is developed in windings that are stationary and arranged in what is called a *three-phase connection*. These windings are in turn connected to six diodes, which are connected as shown in Figure No. 35; this is called a *full-wave bridge*. From an analysis of the arrows on the diodes, it can be seen that ac created in the stator is rectified by the bridge so that dc exists at the output terminals of the alternator. The voltage generated in the three-phase windings is due to the rotating field coil. This rotating field coil causes changes in the magnetic lines of force flowing through the three-phase windings and creates the output voltage of the alternator. Alternators are characterized by longer service life and reliability, higher charging rates at engine idle, and more power in a smaller package.

The alternator output curve illustrated shows that an alternator can produce a certain maximum current for any one design, and the alternator can operate at this maximum output continuously without any danger of burnout. Therefore, there is no need to have any type of control on the maximum charging current of an alternator system.

As can be seen from the connection of the six diodes, the diodes prevent the battery from ever allowing a current to flow from the battery to the alternator, so there is no need for a cutout.

The voltage output of the alternator depends on the speed of rotation and the magnetic field strength generated by the rotor assembly. Voltage regulation is attained by controlling the current flow through the alternator field.

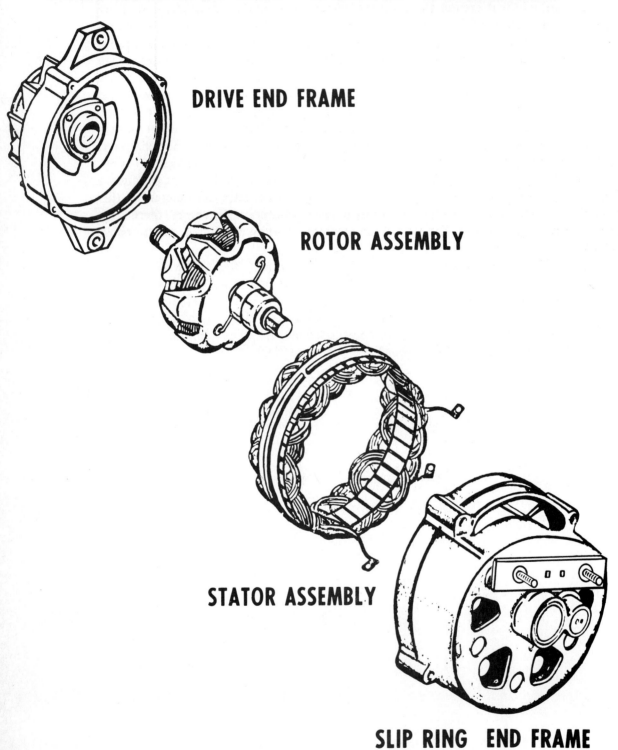

Figure No. 36

ALTERNATOR COMPONENTS

ALTERNATOR COMPONENTS

The alternator is composed of a rotor assembly, a stator assembly, and two end frame assemblies, one at the drive end of the alternator and one at the slip ring end. The rotor assembly is composed of a field coil made of many turns of wire wound over an iron core, which is contained between two iron segments with interlacing fingers. These fingers serve as magnetic poles. This assembly is press mounted on a steel shaft, which turns in prelubricated antifriction bearings. Two slip rings are mounted on one end of the shaft. Each end of the field coil winding is connected to one of the slip rings. A brush rides on each slip ring. These brushes conduct battery current to the rotor winding to create the magnetic field required for voltage generation.

The stator assembly is composed of a laminated iron frame and three sets of windings wound into slots in the frame. The manner in which these windings are wound and connected makes the alternator a three-phase unit. In the most commonly used connection, one end of all three windings is connected together, while the other end of each winding is connected to a pair of diodes, one positive and one negative.

When assembled, a small air gap is present between the rotor poles and the stator to keep the magnetic field lines of force as strong as possible. As the rotor spins, the alternator north and south poles of the rotor fingers pass each loop in the stator windings, inducing an alternating voltage in the windings. The alternating current produced is then rectified by the diodes.

The slip ring end frame contains six diodes, which are electrical rectifying devices. The diodes, three negative and three positive, act as one-way valves, permitting current to pass freely in one direction but not in the other. By their combined action, the alternating current generated is rectified to direct current.

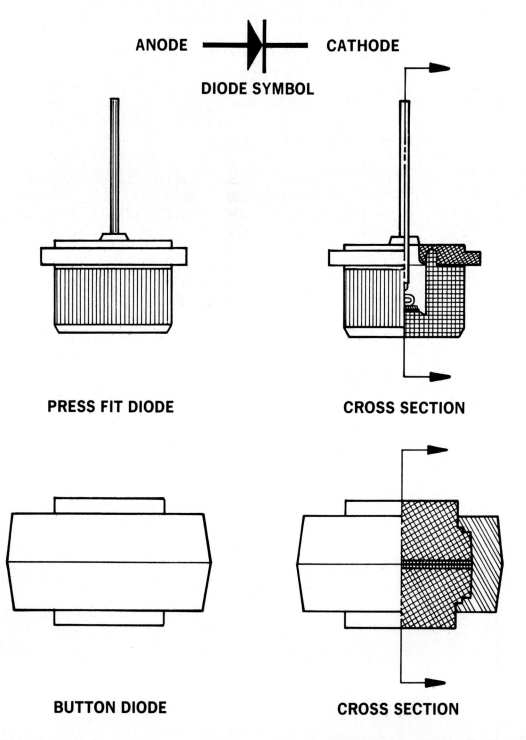

DIODES

As previously stated, the diode is a current-rectifying device. It serves as a one-way electrical check valve that permits current to flow readily in one direction, but stops its flow in a reverse direction.

The diode symbol (Figure No. 37) is an arrow indicating the direction of current flow allowed by the diode. The bar indicates a one-way "gate" or block to current flowing in the opposite direction.

In the press-fit type, the cross-sectional view illustrates the position of the silicon wafer in the bottom of the diode case. The case is made of rather heavy metal to serve both as protection for the rather brittle silicon wafer and to effectively dissipate the heat induced by the current flowing through the diode. The case is tightly sealed during manufacturing to prevent the entrance of any contaminants into the diode, which would result in degradation and shortened life of the unit. Contaminants can be drawn into any unit that is not properly sealed and operates at varying temperatures, since it "breathes" as it heats and cools.

In alternators using press-fit diodes, the negative (case) diodes are frequently pressed into the alternator-grounded end frame, and the positive (case) diodes are pressed into a holder called a *heat sink*. The heat sink is usually made of die-cast aluminum because it possesses high heat-dissipating qualities. It is mounted in, but electrically insulated from, the end frame. The end frame also serves to absorb the heat developed by the passage of current through the diodes.

TRANSISTORS

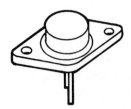 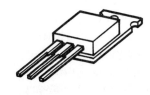 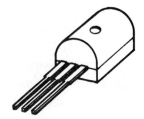

APPEARANCE

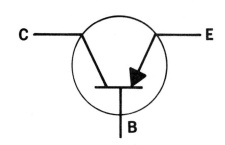 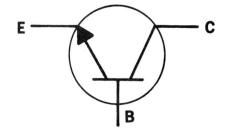

PNP SYMBOL   NPN SYMBOL

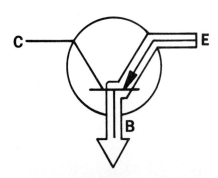 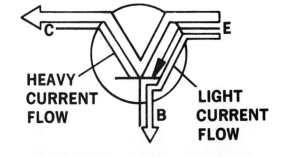

EMITTER-BASE CIRCUIT   EMITTER-COLLECTOR CIRCUIT

TRANSISTORS

A transistor is a device which acts as an electrical switch but has no moving parts. Current flow through the transistor is controlled by the base current.

A transistor is made up of a pure silicon chip that is *doped* in sections with impurities and placed in a container. The sections are referred to as *N*- or *P*-type material. The arrangement of the sections determines the transistor's polarity. The transistor may be known as a *PNP* transistor or as an *NPN* transistor according to the positions of the wafers. The middle wafer serves as the base and therefore dictates the polarity.

The transistor symbol is illustrated in Figure No. 38. The circle represents the container. The letters represent the three elements of which the transistor is composed: *E* is the emitter, *C* is the collector, and *B* is the base. The arrow on the emitter symbol indicates the direction of current flow through the transistor.

There are two important factors relative to the manner in which the transistor works. First, the emitter-collector circuit is the main current-carrying circuit. Second, current flow through the emitter-collector circuit is possible *only* when there is current flow through the emitter-base circuit. Although the current flow in the base circuit may be only a fraction of the current flow in the collector circuit, the collector circuit cannot exist without the base circuit. It follows, then, that an interruption of the light current flow in the base circuit will cause a stoppage in the heavy current flow in the collector circuit. In this manner the base circuit "triggers" the transistor and turns it *on* or *off*.

Figure No. 39

ALTERNATOR WITH DIODE TRIO

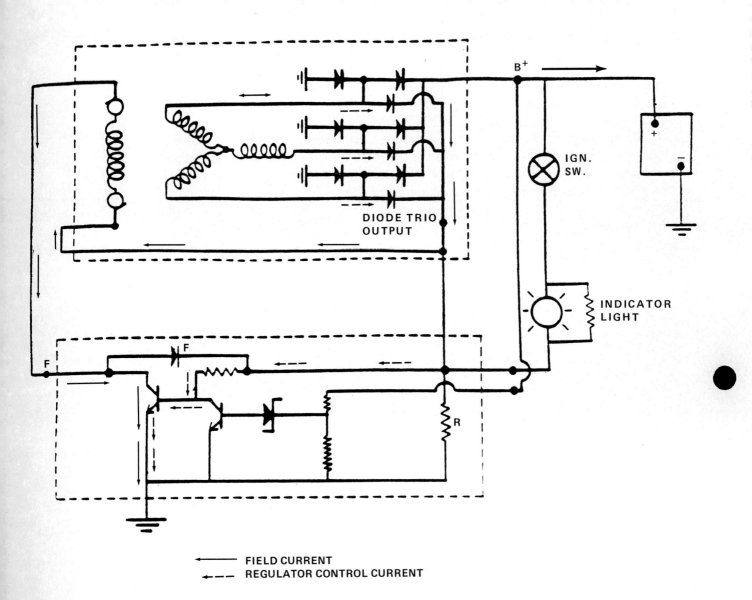

ALTERNATORS WITH A DIODE TRIO

The diode trio simplifies wiring, as it automatically provides field current when the alternator is functioning and automatically turns the field current off when the alternator is not operating.

Figure No. 39 illustrates the circuitry used by General Motors. The indicator light current provides a small but sufficient current to the alternator field to enable the alternator to start functioning when the engine is started. As soon as the alternator starts functioning, the diode trio provides full field current and turns off the indicator light by putting battery voltage at both terminals of the indicator light.

Although the electrical circuitry used by General Motors is complex, it functions as previously described. The resistor *R* allows the indicator light to light up to indicate trouble in the event of an open alternator field circuit, due either to brush-to-slip ring contact problems or an open field coil. The battery-to-regulator lead carries practically no current and therefore provides the regulator with an accurate measure of the system voltage at the battery.

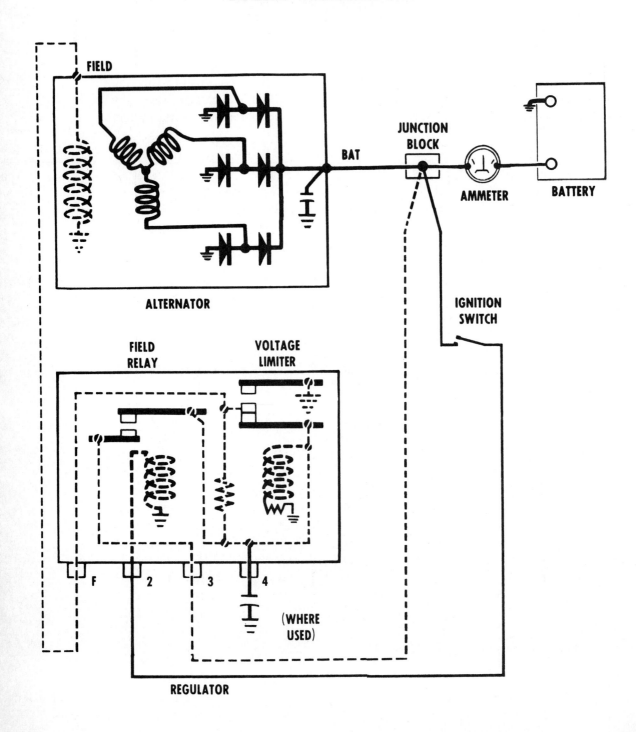

Figure No. 40

AC CHARGING SYSTEM
With Ammeter

AC CHARGING SYSTEM (WITH AMMETER)

A typical ac charging system equipped with an ammeter is illustrated in Figure No. 40. When the ignition switch is closed, a connection is established between its battery and ignition terminals. Battery current now flows to terminal 2 of the regulator, through the field relay shunt winding to ground, and back to the battery through the ground circuit.

The magnetic field created around the field relay core attracts the relay armature, thereby closing the relay points. Battery current now flows to terminal 3 of the regulator, across the field relay points, across the lower voltage regulator points, from the field terminal on the regulator to the field terminal on the alternator, through the field coil to ground, and back to the battery. Some current also flows through the shunt winding on the voltage coil. With the rotor field coil energized, the alternator is ready to produce current as soon as the rotor is turned.

With the field relay closed, current is supplied directly to the alternator field coil from the battery instead of through the ignition switch and ignition primary resistance wire.

When the engine is started, the alternator rotor spins, and the magnetism created in the field coil by the field current induces an alternating current in the stator windings. The diodes rectify the alternating current generated into direct current, as previously explained.

As the vehicle is put in motion, the alternator speed is increased, resulting in a greater voltage being induced on the current flow in the field circuit. Since the shunt winding on the voltage coil is also subjected to this increased voltage, a greater magnetic field is created around the voltage coil. This strong field attracts the voltage regulator armature, causing the lower contact points to separate.

Field current must now flow through the resistor on its way to the field coil. Field current is reduced from approximately 2 amperes to about ¾ ampere by the resistor. The reduced field current results in an immediate reduction in alternator output, with an associated drop in voltage applied to the voltage regulator coil. The voltage regulator armature spring closes the lower contacts, thereby reestablishing full field current flow. This cycling action of inserting and removing the resistor from the field circuit limits the voltage developed by the alternator to a safe value.

If the vehicle is driven at high speed and the accessory and battery demands are low, a higher voltage of 0.1 to 0.3 volt is induced on the shunt coil of the voltage limiter. This results in the upper armature being attracted to the relay core, thereby closing the upper contacts. At this time, both ends of the field coil are grounded, with the result that there is no current flow through the coil. With a "dead" field coil, alternator voltage decreases, permitting the upper contact points to open. Field current now flows through the resistor to the field coil. As the voltage again increases, the upper relay contacts are again closed. The cycling that takes place limits the field current between ¾ ampere and no current at all. By this action, alternator output is safely limited regardless of how fast the vehicle may be driven or how long the speed is sustained.

The function of the condenser used in the alternator is to dampen the high-voltage surges developed in the stator windings or any transient high-voltage impulses generated anywhere in the charging system. The condenser also serves as a noise-suppression unit.

AC CHARGING SYSTEM
With Indicator Lamp

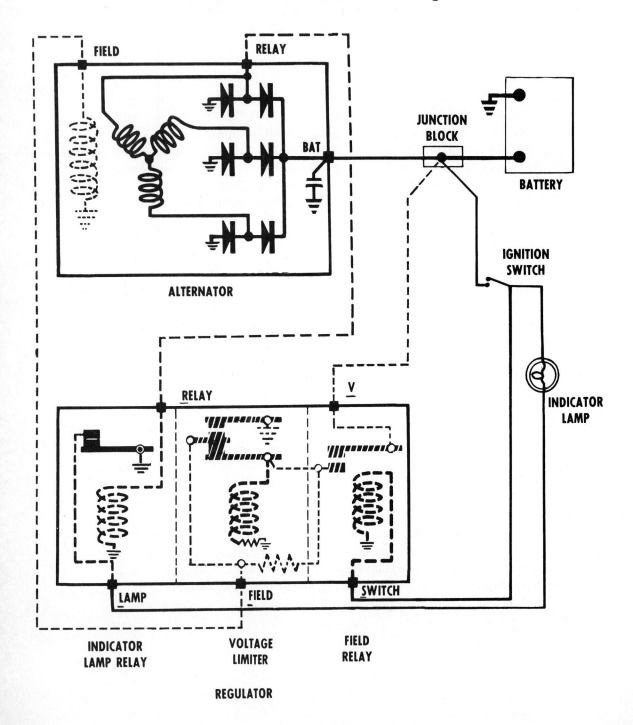

AC CHARGING SYSTEM (WITH INDICATOR LAMP)

A typical ac charging system equipped with an indicator lamp is illustrated in Figure No. 41. When the ignition switch is closed, battery current flows through the indicator lamp to the L terminal on the regulator, across the closed relay contacts, and to ground. This completes the indicator lamp relay circuit, permitting the lamp to light.

Also energized by the closing of the ignition switch is the field relay. Battery current flows from the switch to the SW terminal on the regulator, through the field relay voltage coil to ground, and back to the battery. The magnetic field created around the field relay core attracts the relay armature, closing the relay points. Battery current now flows from the junction block terminal to the V terminal on the regulator, across the field relay points, across the lower voltage regulator points, from the F terminal on the regulator to the alternator, through the field coil to ground, and back to the battery.

When the engine is started, the alternator rotor spins, and the magnetism created in the field coil by the field current induces an alternating current in the stator windings. The diodes rectify the alternating current generated into direct current, as previously explained.

As soon as the alternator starts operating, current flows from the alternator *relay* terminal to the R terminal on the regulator, through the voltage coil on the indicator lamp relay to ground, and back to the alternator. This current flow magnetizes the light relay core, which attracts the relay armature, thereby opening the relay points. This opens the indicator lamp circuit, and the lamp goes out. This circuit arrangement provides a light which serves as a warning when lit with the engine running that trouble exists in the charging system.

As the vehicle is put in motion, the alternator speed is increased, resulting in a greater voltage being induced on the current flow in the field circuit. Since the shunt winding on the voltage coil is also subjected to this increased voltage, a greater magnetic field is created around the voltage coil. This strong field attracts the voltage regulator armature, causing the lower contact points to separate. Field current must now flow through the resistor on its way to the field coil. Field current is reduced from approximately 2 amperes to about ¾ ampere by the resistor. The reduced field current results in an immediate reduction in alternator output, with an associated drop in voltage applied to the voltage regulator coil. The voltage regulator armature spring closes the lower contacts, thereby reestablishing full field current flow. This cycling action of inserting and removing the resistor from the field circuit limits the voltage developed by the alternator to a safe value.

If the vehicle is driven at high speed and the accessory and battery damands are low, a higher voltage of 0.1 to 0.3 volt is induced on the shunt coil of the voltage limiter. This results in the upper armature being attracted to the relay core, thereby closing the upper contacts. At this time, both ends of the field coil are grounded, with the result that there is no current flow through the coil. With a dead field coil, alternator voltage decreases, permitting the upper contact points to open. Field current now flows through the resistor to the field coil. As the voltage again increases, the upper relay contacts are again closed. The cycling that takes place limits the field current between ¾ ampere and no current flow at all. By this action, alternator output is safely limited regardless of how fast the vehicle may be driven or how long the speed is sustained.

© Copyright 1986, Tune-Up Manufacturers Institute

Figure No. 42
AC CHARGING CIRCUIT INDICATOR LAMPS
Indicator lamp with separate lamp relay

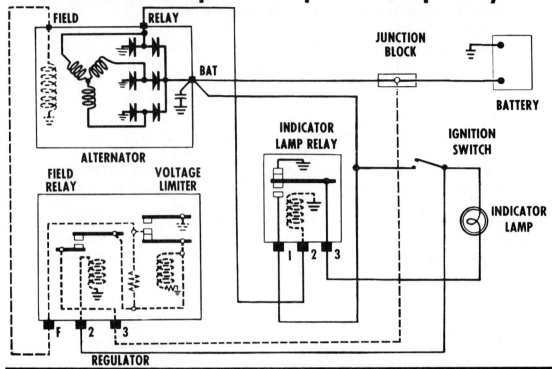

Indicator lamp without lamp relay

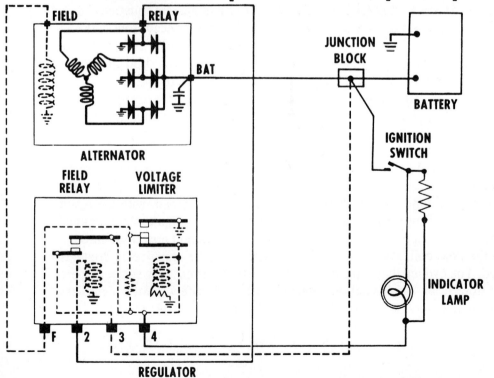

© Copyright 1986, Tune-Up Manufacturers Institute

AC CHARGING CIRCUIT INDICATOR LAMPS

The upper circuit diagram Figure No. 42 illustrates the use of an indicator lamp and a separate indicator lamp relay. When the ignition switch is turned on, battery current flows through the indicator lamp, to indicator lamp relay terminal 3, across the relay armature and the closed upper relay points to ground. This completes the circuit, permitting the indicator lamp to light.

As soon as the engine starts, alternator voltage from the relay terminal is impressed on indicator lamp relay terminal 2. The current flows through the relay winding, creating a magnetic field which attracts the relay armature, pulling it down and closing the lower contacts.

When this occurs, not only is the circuit ground denied to the indicator lamp, but alternator output voltage is now applied to both sides of the lamp at the same time. As a result, current flow stops and the light goes out.

The rest of the charging system functions are as previously explained.

The lower circuit diagram illustrates the use of an indicator lamp without the use of an indicator lamp relay.

When the ignition switch is turned on, battery current flows through the indicator lamp to regulator terminal 4, across the closed lower voltage regulator points, to the regulator F terminal, and to the alternator field coil. This complete circuit permits the indicator lamp to light.

As soon as the engine starts, alternator voltage is impressed on the alternator relay terminal causing current to flow to regulator terminal 2 and through the shunt winding in the field relay. The magnetic field created by this current flow attracts the field relay armature, closing the field relay circuit. System voltage impressed on regulator terminal 3 is now also impressed on terminal 4. This results in system voltage being applied to both sides of the indicator lamp at the same time. With no current flow through the lamp, the light goes out.

In this regulator circuit, it may be said the field relay has a dual function. It not only completes the field circuit directly from the battery to the alternator field, instead of through the ignition switch and primary resistance wire, but it also serves as an indicator light relay.

The rest of the charging system functions are as previously explained.

Figure No. 43

ALTERNATOR VOLTAGE CONTROL

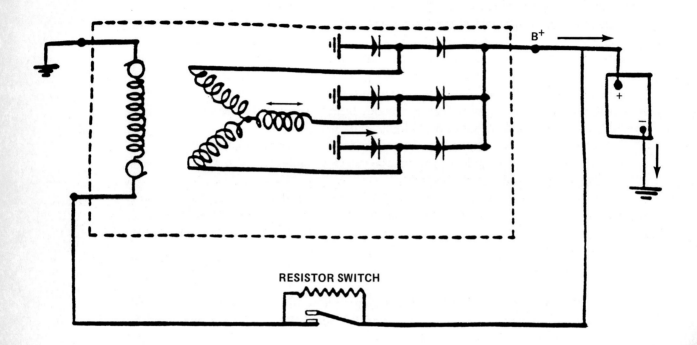

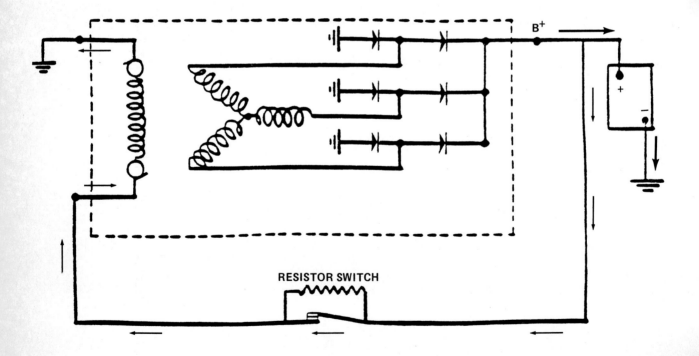

ALTERNATOR VOLTAGE CONTROL

Figure No. 43 illustrates the basic method of voltage control of alternators whether using mechanical or solid-state means. The simple diagram illustrates how the battery is used to provide a voltage source for field current. With the switch closed, maximum field current flows and the alternator generates at the maximum voltage that can be obtained for its rotating speed. In normal operation with full field, the alternator voltage will rise beyond the 14.2 volts that we are looking for in a charging system. In practice, when the voltage rises to 14.3 volts, some device opens the switch shown. When the switch opens, full current must now flow through the resistor shown, and this resistance is now in series with the field coil resistance. According to Ohm's law, the field current will now be reduced, and as long as sufficient resistance is inserted, the field current will drop so that the alternator can no longer develop 14.3 volts, and the alternator voltage will drop. When the voltage decreases to 14.1 volts, the actuating device will now close the switch, thus removing the resistance and causing both field current and the alternator voltage to rise. As soon as it rises to 14.3 volts, the switch opens and the basic cycle is repeated. Thus, in practice, the alternator voltage varies between 14.3 and 14.1 volts at a very rapid rate. The actual opening and closing voltage of the switch varies with the particular type of regulator used, but the basic action is always the same. This voltage variation is extremely rapid, and a voltmeter connected to the alternator will read the average value of 14.2 volts. The actual means of opening and closing the switch, whether it be mechanical or solid state, will be described later.

The internal resistance of a battery varies inversely with temperatures. As temperature increases, the internal resistance decreases. To offset this characteristic, voltage regulators are designed to have temperature compensation. Although the amount of temperature compensation may vary in different designs, in general, a regulator that operates at 14.2 volts at 70°F will operate in the neighborhood of 15 volts in subzero temperatures and around 13.6 volts under hot driving conditions on hot days.

The voltages given here are those that have been used for many years with the lead-acid battery to which water must be added. The maintenance-free type of batteries tolerate about 0.5-volt higher settings, and vehicle manufacturers may opt to use higher settings when using these batteries.

Figure No. 44

VOLTAGE REGULATOR RELAYS

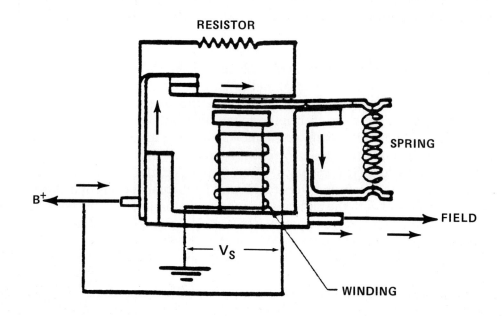

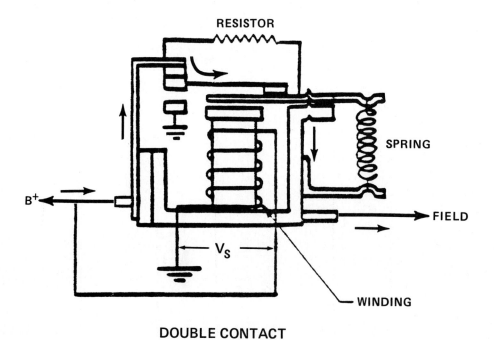

DOUBLE CONTACT

MECHANICAL VOLTAGE REGULATOR

The basic mechanical voltage regulator contains a winding of many turns of small copper wire. This winding creates a magnetic field that creates a pull on the armature plate. The relay spring exerts a pull that is opposite to the magnetic pull and also tends to hold the contacts closed. The contacts themselves are the switch illustrated in Figure No. 44 and carry the alternator field current. The voltage of the alternator sends a current through the relay windings. This current creates magnetic lines of force, which create a pull on the armature that tends to open the points. The relay spring opposes this pull and tends to keep the contacts closed. As soon as the magnetic pull exceeds the spring pull, the contacts will open. Expressing this in mathematical form, we can say

$$\frac{V \text{ system}}{R \text{ winding}} = I_{winding} = \phi_{mag} = \text{pull relay}$$

Simplified: V system = pull relay
Pull relay = pull spring − relay operating point

Therefore,

V system = pull spring

Thus it can be seen that as a system voltage rises the magnetic pull will rise until it exceeds the pull of the relay spring. At this point the contacts open, inserting a resistance in the field circuit and causing the alternator voltage to fall. As the voltage falls, the magnetic pull decreases and the spring pressure causes the contacts to reclose; thus voltage regulation of the alternator is attained.

Due to the high field currents used with alternators, the actual voltage regulator used is called a *double-contact* design. This design uses a relatively low value of resistance across the normally closed contacts, somewhere in the range of 10 to 14 ohms. Under conditions of a high charging rate and low alternator speed, the addition of the 14 ohms into the alternator field circuit is sufficient to limit the alternator output voltage. However, under conditions of high alternator speed and low alternator output, this resistance is not sufficient to limit the rise of alternator voltage. As the alternator voltage rises approximately 0.3 to 0.5 volt, the increased voltage causes a further movement of the armature plate of the relay until the second set of contacts closes. When the second set of contacts closes, both ends of the alternator field are connected to ground, which causes the alternator voltage to fall. As the voltage of the system falls, the relay magnetic pull weakens and the second set of contacts opens, allowing the voltage to rise. Thus the second set of contacts controls the alternator voltage at a slightly higher voltage value than the first set of contacts does, but under conditions of high alternator speed and light load.

The single relay voltage regulator is used on vehicles equipped with an ammeter. The ammeter indicates to the vehicle operator if the system is functioning properly.

MECHANICAL VOLTAGE REGULATOR
With Indicator Lamp

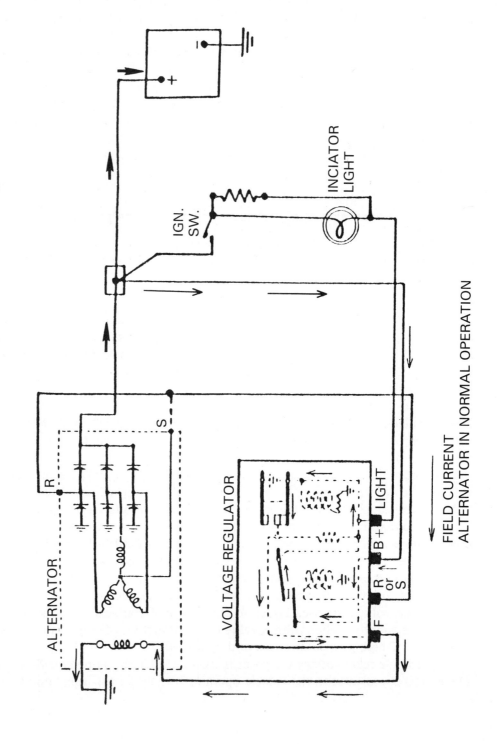

Figure No. 45

MECHANICAL VOLTAGE REGULATOR WITH INDICATOR LAMP

When vehicles are not equipped with an ammeter or with a voltmeter, they are equipped with a dashboard-mounted indicator light to indicate if the system is operating. The voltage regulator used with this system also contains a simple, normally open relay, commonly called a *slave* relay. The slave relay is operated by an *R* or *S* terminal in the alternator, both of which are shown in Figure No. 45. General Motors has used the R terminal, while Ford uses the S connection. When an alternator is not operating, both the R and S terminals will be at 0 volt, and when the alternator is operating the R and S terminals will be at approximately one-half output voltage, or 7 volts. Thus, when this system functions, the 7 volts operates the slave relay to turn off the indicator light.

In Figure No. 45, before the engine starts and the ignition switch is turned on, current flows through the indicator lamp, voltage regulator contacts, and through the field of the alternator. The amount of current that flows through the circuit is sufficient to allow the alternator to build to at least 13 volts when the engine is started. When the alternator operates at 13 volts, the R or S terminal voltage of 6.5 volts closes the slave relay contacts. As can be seen from the schematic, this applies battery voltage to the R terminal, full voltage to the field, and the indicator light goes out because it has battery voltage at both of its terminals.

The resistance in parallel with the indicator lamp allows the charging system to operate if the indicator lamp should burn out.

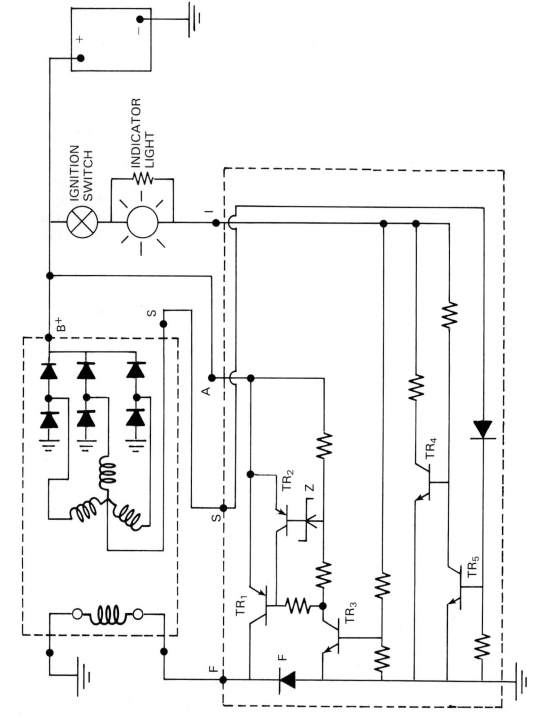

ELECTRONIC (TRANSISTOR) VOLTAGE REGULATOR

Figure No. 46

ELECTRONIC (TRANSISTOR) VOLTAGE REGULATOR

As high-reliability, low-cost transistors became available, it was inevitable that they would be used to provide the voltage-regulator function in alternator charging systems. Electronic voltage regulators function in the same manner as the mechanical voltage regulator in that they have a transistor that when turned on allows full field current to flow, and when turned off interrupts the field current.

Figure No. 46 illustrates the basic circuit of an electronic voltage regulator. TR_1 is connected in series with the alternator field, and when the ignition switch is closed, it is *on*, and base–emitter current is supplied through resistor A. Thus full field current flows through the alternator TR_1 to ground, and when the engine is started the charging system voltage rises until the voltage at point B is greater than the breakdown voltage of zenor diode Z and the base–emitter voltage drop of TR_2. When this occurs, TR_2 turns *on*. The collector–emitter voltage of TR_2 is less than the base–emitter voltage of TR_1, and the normal base–emitter drive current of TR_1 is shunted through TR_2, thus turning TR_1 off. With the alternator field current off, the system voltage drops, and point B voltage drops until TR_2 turns off. This turns TR_1 on and the charging system voltage rises, and the switching cycle continues at a very rapid rate, holding the system voltage constant.

With the circuit shown, the alternator field current is turned on when the engine is started and off when the engine is stopped by the ignition switch. The diode F is used to prevent voltage spikes that can be created by field current switching from damaging TR_1.

This arrangement uses two isolated brushes, and one side of the field is connected to the battery and the other side is connected to ground through the transistor TR_1. This type of circuit uses an *NPN* power transistor. The isolated brush-type alternators are used by Chrysler and General Motors with solid-state regulators.

Solid-State Regulators Used by Ford

The two schematics in Figure No. 46 illustrate the circuitry used by Ford on vehicles using indicator lights and those that are equipped with ammeters. On a vehicle equipped with an indicator light, when the ignition switch is turned on with the engine not running, the indicator light is turned on and also the field current. When the engine is started, as the alternator builds to normal voltage, the S terminal voltage rises to one-half of the alternator output voltage. This action turns on TR_5, which turns off transistor TR_4 to turn off the indicator light.

On those vehicles that are equipped with an ammeter, when the ignition switch is turned on, TR_3 is turned on, turning on field current. Normal action of the charging system is then monitored by the ammeter.

In operation, normal voltage-regulator action occurs as previously explained with transistors TR_1 and TR_2. With this system, Ford uses an alternator with one brush grounded, and the TR_1 and TR_2 transistors are *PNP* type. It should be apparent that transistors TR_4 and TR_5 and the associated circuitry replace the function of the slave relay used on the mechanical regulator.

Figure No. 47

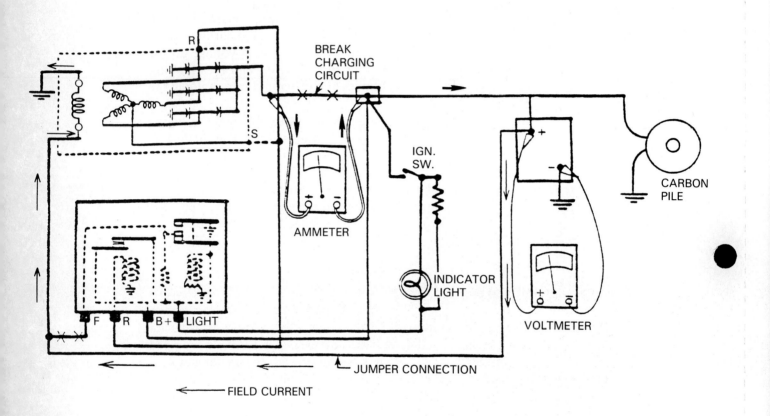

TEST FOR ALTERNATORS WITH MECHANICAL REGULATORS

Problems with charging systems fall into two classes: low-voltage or high-voltage conditions. With test equipment connected as shown in Figure No. 47, and the alternator rotating at high speed, if the system cannot produce the alternator's rated output at normal system voltage, then either the alternator is malfunctioning or the control circuit is not allowing full field current. To determine which is the case, a jumper wire is used to connect battery + potential to the isolated brush (all these alternators have one brush grounded), allowing full field current to flow. If under these conditions the alternator produces rated output, then the regulator is faulty and needs to be serviced or replaced. On the other hand, if with full field current the alternator cannot produce full output, then the alternator is faulty. Some common causes of this latter case are an open or shorted diode, poor brush-to-slip ring contact, or an open field. When doing these tests, always disconnect the field terminal at the regulator because if the alternator is functioning the regulator can be destroyed.

If a high-voltage condition exists, it is evident that the alternator is operating with too high a field current. Troubleshooting this condition is done by removing the field lead at the regulator. If the alternator now stops functioning, then the regulator was providing too much field current and needs service. Common causes of such malfunctioning are sticking regulator contacts, high relay spring tension, or a poor regulator ground. If the voltage and output remain high, then some wiring malfunction is indicated as full field current continues when the regulator is removed. Such a wiring malfunction is not common.

The carbon pile is used to load the system to determine if the alternator can produce rated output. The loading provided by the carbon pile also prevents the system voltage from rising too high under maximum output conditions so as to protect other sensitive items in the vehicle.

Figure No. 48

TESTING ALTERNATORS WITH ELECTRONIC REGULATORS

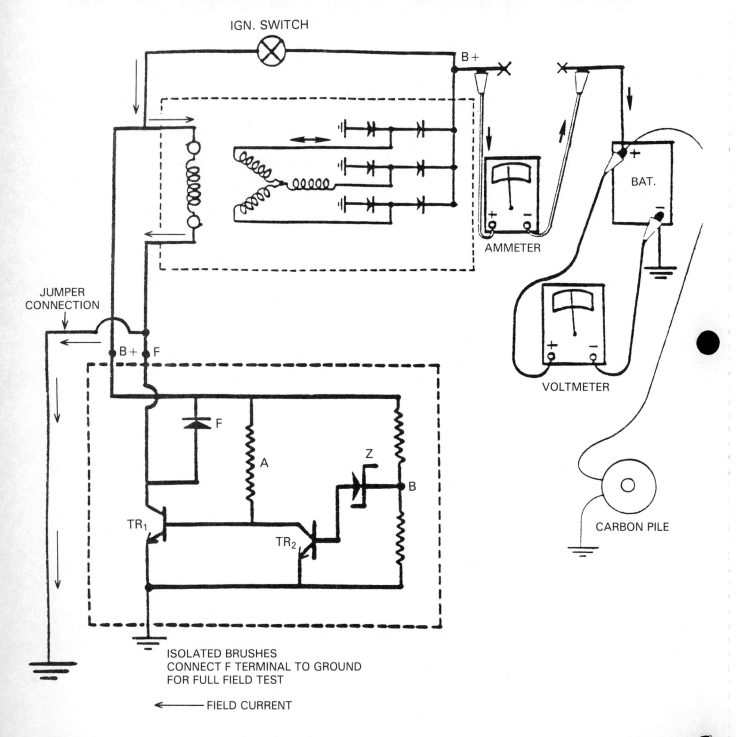

TEST FOR ALTERNATORS WITH ELECTRONIC REGULATORS

The testing of charging systems with electronic regulators is essentially the same as for those with mechanical controls. Field current is increased during low battery voltage periods and reduced as battery voltage approaches charging system output.

In the troubleshooting procedure in which full field current is applied to the alternator, it must first be determined whether the alternator has two isolated brushes or one brush grounded. In the case of the alternator with two isolated brushes, the one brush that is normally grounded through the regulator is connected to ground through the use of a jumper wire, thus bypassing the regulator. This is the case for Chrysler systems. General Motors also provides a means of applying full field current by bypassing the regulator through the use of a small test hole in the rear alternator end cover. When a small screwdriver or similar tool is inserted in hole, the field becomes effectively grounded, applying full field current to the alternator.

Ford-type alternators that have a regulator mounted externally to the alternator have one brush grounded internally in the alternator. Applying full field current is done by a jumper wire from BAT+ to the field terminal that is connected to the regulator.

In the case of electronic voltage regulators, it is not necessary to disconnect the regulator from the alternator when making a full field connection.

ALTERNATOR TESTING
GENERAL MOTORS DELCOTRON

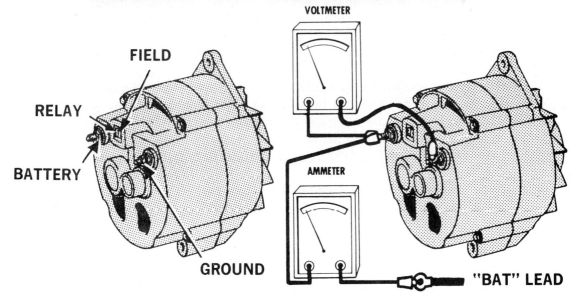

TERMINAL IDENTIFICATION

DELCOTRON TEST HOOK-UP

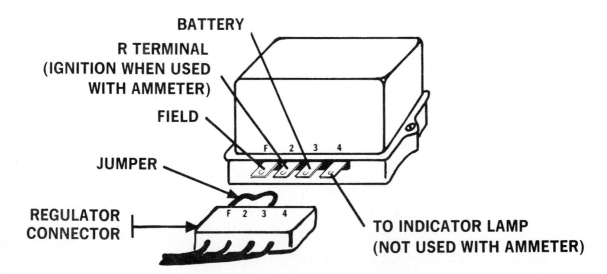

TWO-UNIT REGULATOR

ALTERNATOR TESTING—GENERAL MOTORS DELCOTRON

This procedure is for testing present models of Delcotron units using a two-unit regulator, which consists of a voltage regulator and a field relay. The charging system may employ either an indicator light or an ammeter.

Delcotron Output Test

For 1972 and later models, remove air conditioner-heater motor fuse.

1. Check and adjust drive-belt tension as required.
2. Disconnect battery ground cable.
3. Disconnect lead from Delcotron BAT terminal and connect ammeter between terminal and disconnected lead.
4. Reconnect battery ground cable.
5. Connect carbon pile rheostat across battery. Connect tachometer.
6. Connect voltmeter across Delcotron from BAT terminal to GRD terminal or across battery.
7. Turn on all electrical accessories, set parking brake firmly, start and idle engine with transmission selector in *drive* position. Delcotron output should be at least 10 amperes.
8. Shift selector to *park* position and increase engine speed to 1500 rpm. Ampere output should be as specified. Adjust load rheostat, if required, to obtain desired output.
   Note: In the absence of a carbon pile rheostat, turning the headlights on high beam usually supplies the necessary load.

If output is less than specified, supply field current directly to Delcotron by either of two methods, as follows:
Unplug Delcotron connector and connect jumper lead from F terminal to BAT terminal in connector or connect jumper lead from Delcotron F terminal to BAT (3) terminal. Repeat test in step 8.
Caution: Increase engine speed slowly to prevent voltage from exceeding maximum specified limit.

1. If output remains low, Delcotron is defective.
2. If output rises to normal output, regulator or connecting wiring is at fault.
3. Stop engine. Remove jumper lead and reconnect harness. Install fuse.

Voltage Regulator Test

1. Install ambient temperature gauge on regulator cover.
2. With all electrical loads *off*, run engine for 15 minutes at about 1500 rpm.

3. Note ammeter reading. For accurate voltage check, ammeter should read between 3 and 10 amperes. If ammeter reading is still high after 15-minute run, substitute a fully charged battery and proceed with test.

   Note: A ¼-ohm resistor in series with the ammeter will simulate a fully charged battery and will permit an uninterrupted test.

4. Momentarily increase engine speed to about 2000 rpm and note readings on voltmeter and temperature gauge. Check readings against specifications for voltage setting at regulator ambient temperature. Make note of amount of adjustment required to place setting in specified limits.

   a. Remove regulator cover. *Be careful* to lift cover *straight up.* If field relay or voltage coil are touched by cover, the resulting arc may ruin the regulator. Voltage reading will change somewhat when cover is removed.

   b. Increase or decrease voltage setting the amount previously noted. Always make final adjustment by slightly increasing the spring tension.

   c. Carefully replace cover and recheck voltage setting.

The spread in the voltage settings is provided to permit tailoring the setting to the specific requirements of the vehicle being tested. In any event, however, the final setting must be within specified range.

Notes

Figure No. 50

ALTERNATOR TESTING

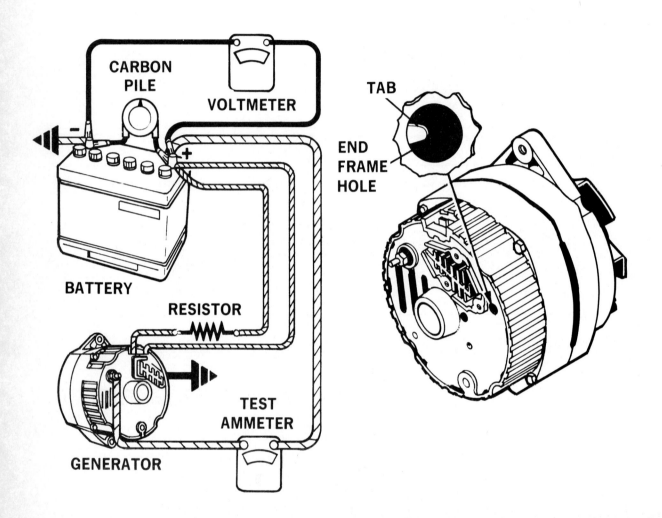

Full field current can be obtained by inserting a screwdriver through the end frame hole and grounding the metal tab to the frame.

TESTING THE DELCOTRON WITH INTERNAL REGULATOR

ALTERNATOR TESTING—GENERAL MOTORS DELCOTRON WITH INTEGRAL REGULATOR

This procedure is for testing Delcotron units that use an integral (internal) regulator and incorporate either an indicator light or an ammeter.

Delcotron Output Test

Important: When testing, all accessories must be shut off and the blower motor lead disconnected.

1. Check and adjust drive-belt tension as required.
2. As a safety precaution, disconnect battery ground cable while connecting test instruments.
3. Disconnect the lead from the BAT terminal, and connect a test ammeter (0- to 75-amp range or higher) between the disconnected wire and the BAT terminal.
4. Reconnect battery ground cable.
5. Operate engine at recommended test speed (typically 1500 to 2000 rpm).
6. Using a screwdriver, momentarily ground the tab to the Delcotron housing (see Figure No. 50) and observe the test ammeter. The reading should equal the specified output current ± 10 percent. If so, the Delcotron is considered good. If not, the unit must be removed for bench testing and servicing.

Note: Do not ground the tab any longer than is necessary to obtain an ammeter reading.

Delcotron Voltage Regulator Test

1. Connect a voltmeter (0- to 16-volt range) across the battery terminals. Leave ammeter connected as before.
2. With all electrical loads off, run engine for about 15 minutes at a fast idle. This is to normalize the system and, if necessary, partially recharge the battery.
3. At the specified test speed (usually 1500 rpm), note the ammeter reading. If it is less than 10 amperes, check the voltmeter reading. This will be the voltage regulator setting, which should be within the specified limits (for example, 14.0 ± 0.5 volts). The regulator is nonadjustable. If the ammeter shows more than 10 amperes, it indicates that the battery is too discharged for accurate voltage regulator testing. Install a fully charged battery and repeat the test. *Note:* Some charging system testers have provision for inserting a ¼-ohm resistor in the charging system to simulate a fully charged battery. This allows accurate tests of the voltage regulator with a discharged battery. Follow the instrument maker's prescribed test procedure.
4. An out-of-tolerance voltage regulator must be replaced. Since the regulator is internal, the Delcotron must be removed from the vehicle.

© Copyright 1986, Tune-Up Manufacturers Institute

Figure No. 51

ALTERNATOR TESTING
CHRYSLER CORPORATION

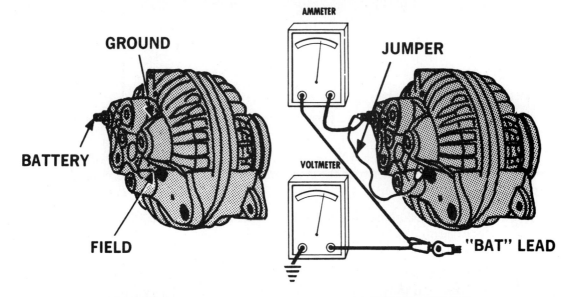

TERMINAL IDENTIFICATION

ALTERNATOR TEST HOOK-UP

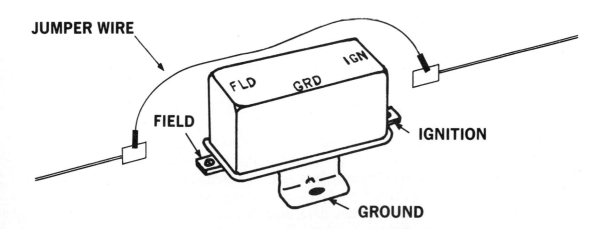

SINGLE-UNIT REGULATOR

Full field current can be obtained by connecting a jumper wire between the field and ignition terminals on mechanical regulators.

ALTERNATOR TESTING—CHRYSLER CORPORATION I

The regulator used on Chrysler Corporation cars contains only a voltage-limiting relay. Chrysler recommends that a fully charged battery be installed in the vehicle before charging system tests are conducted if the vehicle's battery is in a state of undercharge. Also make certain the alternator drive belt is properly tensioned.

Alternator Output Test

For 1972 and later models, remove air conditioner-heater motor fuse.

1. Connect test ammeter between alternator BAT terminal and disconnected wire.
2. Connect jumper lead from alternator FIELD terminal to BAT terminal.
3. Connect voltmeter between disconnected BAT terminal lead and a good ground.
4. Connect tachometer.
5. Connect a carbon pile rheostat across the battery. Place rheostat in OPEN or OFF position.
6. Set engine speed to 1250 rpm. Adjust carbon pile to control alternator output voltage at 15.0 volts.
7. Observe ammeter for current output and compare reading to specifications. Output reading should be as specified with ± 3 amperes tolerance.
8. Stop engine. Remove jumper lead and rheostat. Install fuse.

Voltage Regulator Test

The regulator is checked by performing two tests. The first test checks the regulator's ability to maintain a specified voltage at low vehicle speed with heavy electrical loads. The second test checks the ability of the regulator to maintain voltage control at high vehicle speeds with light electrical loads.

First Test (Upper Contacts)

1. Install ambient temperature gauge on regulator cover.
2. Set engine speed to 1250 rpm. Turn on lights and electrical accessories to obtain a 15-ampere charge rate. Operate engine for 15 minutes at this speed and load to normalize charging system temperature.
3. After 15 minutes, reset engine speed to 1250 rpm if required; readjust load to 15 amperes, as required.
4. Observe voltmeter and temperature readings and compare readings to specifications.

© Copyright 1986, Tune-Up Manufacturers Institute

Second Test (Lower Contacts)

5. Increase engine speed to 2200 rpm.
6. Turn off all electrical accessories and observe ammeter and voltmeter.

Amperage should decrease. Amperage outlet should be 5 amperes or less. The voltage should increase at less 0.2 volt but not more than 0.7 volt.

A regulator that does not conform to these specifications must be adjusted. It may be necessary to reset the armature air gap and contact point spacing. Readjust the voltage limiter by bending the spring hanger down to increase the voltage setting or up to decrease the setting. A regulator that cannot be brought into the specified operating range by adjustment must be replaced.

Notes

Figure No. 52

ALTERNATOR TESTING

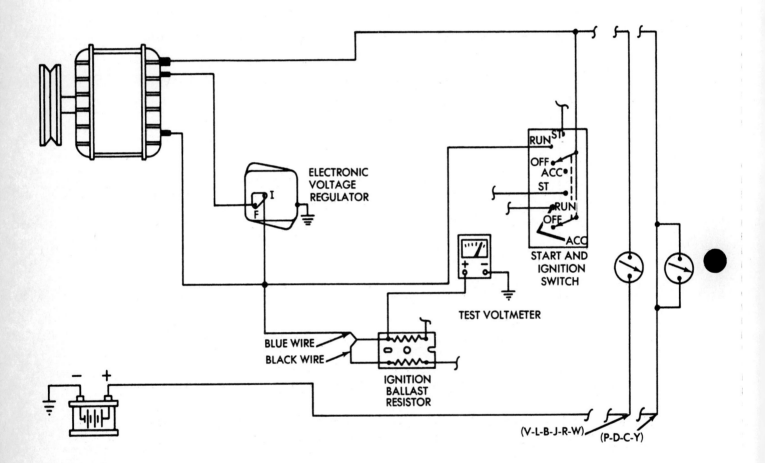

TESTING THE CHRYSLER ELECTRONIC VOLTAGE REGULATOR

ALTERNATOR TESTING—CHRYSLER CORPORATION II

This procedure is for use with systems incorporating a solid-state voltage regulator. Be sure the battery is near full charge before testing. Also check the drive-belt tension and make certain the blower motor lead is disconnected. Turn all accessories off.

Alternator Output Test

1. Disconnect the lead from the alternator battery terminal, and connect a test ammeter (0- to 75-amp range or higher) in series with the disconnected wire and the battery terminal.
2. Connect a voltmeter (0- to 16-volt range) to the battery terminal and ground.
3. Connect a carbon pile across the battery terminals. Be sure the pile is *off*.
4. Disconnect the green field wire from its terminal at the alternator. Set it aside.
5. Connect a jumper lead from this field terminal on the alternator to ground.
6. Start the engine and let it *idle*.
7. Increase the speed while slowly turning in the carbon pile until 1250 rpm and 15.0 volts is obtained. The carbon pile will control the voltage, which should never be allowed to exceed 16.0 volts.
8. At the above speed and voltage, note the ammeter reading. This should be within the specified range. If not, remove the alternator for bench testing.
9. Reduce speed, turn off the carbon pile, and remove jumper lead. Reconnect field lead.

Voltage Regulator Test (Solid-State Unit)

1. Be sure battery is adequately charged.
2. Connect the voltage meter to the battery side of the ignition ballast resistor and ground. Be sure all accessories are off.
3. Run the engine at 1250 rpm and note voltmeter reading. Compare with specified voltage regulator range (typically, 13.8 and 14.4 at 80°F).
4. If the voltage reading is high, check for a poor regulator ground before replacing the regulator. If the reading is low, replace the regulator. The regulator is nonadjustable.
5. Stop engine and remove test instruments.

Figure No. 53

ALTERNATOR TESTING
FORD MOTOR COMPANY

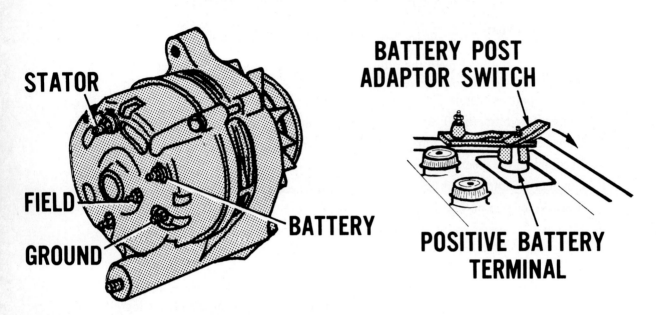

TERMINAL IDENTIFICATION

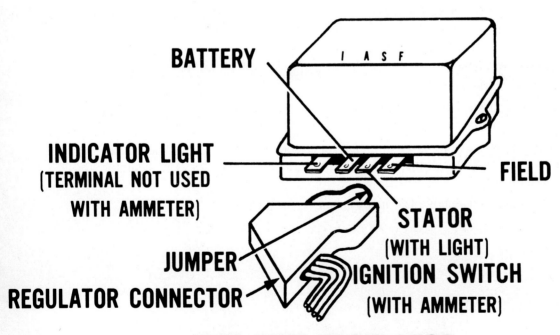

TWO-UNIT REGULATOR

ALTERNATOR TESTING—FORD MOTOR COMPANY

Regulators used by Ford Motors contain two units, a voltage limiter and a field relay. The charging system may employ either an indicator light or an ammeter.

Alternator Output Test

For 1972 and later models, remove air conditioner-heater motor fuse.

1. Remove ground and insulated battery cable clamps from battery.
2. Install battery post adapter switch on positive battery post. Place switch in *open* position. Connect insulated battery cable to battery post adapter.
3. Disconnect plug connector at regulator and connect a jumper lead between the A (battery) terminal and the F (field) terminal of the connector.
4. Connect the tester ammeter leads to the battery post adapter switch, observing polarity. Turn tester control knob to the *direct* position.
5. Reconnect battery ground cable to battery negative post.
6. Connect voltmeter leads across battery, observing polarity.
7. Connect tester ground lead to battery negative post.
8. Observe ammeter. A reverse current flow of from 2 to 3 amperes (field current) should be indicated.
9. Turn tester control knob to *load* position.
10. Close battery post adapter switch and start engine. Open switch. *Note:* The battery post adapter switch is closed only during engine starting operations. It is open at all times tests are being conducted.
11. Slowly increase engine speed to about 2500 rpm while slowly rotating tester control knob to limit alternator output voltage to 15.0 volts.
12. Observe ammeter for alternator output and compare reading to specifications.
13. Stop engine. Remove jumper lead and reconnect connector. Install fuse.

Voltage Regulator Test

Leave battery post adapter switch and tester ammeter and voltmeter leads connected as in alternator output test. All light and electrical accessories must be turned *off*.

1. Install ambient temperature gauge to regulator cover.
2. Close battery switch, start engine, open switch, and set engine speed at 2000 rpm.
3. Rotate tester control knob to the ¼-ohm position. Ammeter should indicate about a 2-ampere current flow.
4. Observe voltmeter and regulator ambient temperature. Compare readings to specifications. A voltage reading out of the allowable range of the specifications should be adjusted to a setting midway in the range.

© Copyright 1986, Tune-Up Manufacturers Institute

A voltage reading within the limits of the specifications, but with either an undercharged or overcharged battery condition, may be tailored to the individual vehicle's requirements by raising or lowering the voltage setting slightly to compensate for the condition. In any event, the final setting must be within the range of the specification.

On-the-Vehicle Diode Test

A low reading during the alternator output test may be caused by one or more defective diodes. This procedure provides a way of testing these alternator diodes without removing the alternator. Only a voltmeter is required.

1. Disconnect the electric choke lead, if used.
2. Disconnect the voltage regulator wiring plug and connect a jumper between the A and the F terminals of this plug (the same connections used for the output test).
3. Connect a voltmeter (0- to 16-volt range) across the battery terminals.
4. Operate the engine at idle speed.
5. Record the voltmeter reading.
6. Move the positive voltmeter lead to the S terminal of the regulator plug and note the voltage reading:
   a. If it is one-half of the reading in step 5, the diodes are good.
   b. If it is more than 1.5 volts above or below one-half of the reading in step 5, one or more defective diodes are indicated.
7. Be sure to reconnect the electric choke (if so equipped) after testing.

ALTERNATOR TESTING FACTORS

Failure of the alternator charging system to function normally is revealed by an indicator lamp that does not light when the ignition switch is turned on, by a lamp that stays lit after the engine starts running, by a slower than normal cranking speed, or by a battery being in a state of undercharge or overcharge.

An alternator with a faulty diode can put out enough current to supply the ignition system demands and yet be incapable of keeping the battery fully charged, especially when the lights and accessories are used.

When an alternator diode is defective, an obvious indication is a whine or hum with the engine idling or operating at low speed. Since the alternator is a three-phase machine, when one diode is defective the machine is out of phase, soundwise, resulting in a whine. This condition is usually the result of a shorted diode. When a diode is open, the condition is generally indicated by noisy operation of the alternator. This noise is caused by the physical unbalance of the unit, which has been created by the electrical unbalance, so to speak. If this condition is allowed to persist, the antifriction rotor shaft bearings in the diode end frame may be damaged.

By far, the greatest percentage of alternator trouble is diode trouble. But usually this trouble is created by improper test procedures, reverse current connections, removing alternator leads while the alternator is in operation, reversing battery connections, and other abuses.

When an alternator charging system complaint is expressed, a few checks and tests should be made before the alternator is condemned or disassembled.

1. Check the tension of the drive belt and inspect its condition. Stretched, frayed, or oil-wetted belts should be replaced. The smaller alternator drive pulley has less wraparound drive-belt action, making belt tension particularly critical. Further, since the alternator has an output even at idle, it is possible for the belt to be slipping at idle speed. Care must be exercised when adjusting belt tension so that the aluminum alternator housing is not crushed by the pry bar. Rest the bar only on the heavy front section of the housing.
2. Test the condition of the battery. A sulfated or internally defective battery will resist being charged even when the charging system is functioning normally.
3. Excessive resistance in the charging circuit can cause a lower-than-normal charge rate and result in a discharged battery. To isolate the point of high resistance with your test equipment, test both the insulated circuit and the ground circuit.
4. Make an alternator field current draw test and an output test.
5. Test the regulator. A malfunctioning field relay may be restricting field current, thereby reducing alternator output. A low-voltage regulator setting can also be responsible for an undercharged battery condition.

If the above checks and tests do not reveal any defective conditions, complete tests of the alternator should be conducted before the unit is disassembled.

Starting with the 1972 continuous blow models, be sure the heater-air conditioner blower motor is disconnected, along with all the other electrical accessories, before conducting an alternator output test. Pulling the air conditioner fuse is a quick way to cut the power to the blower motor.

To avoid the hazard of instant, accidental fogging of the windshield, the blower motor operates continuously at low speed unless regulated otherwise. Failure to disconnect the blower motor before testing will result in a lower than specified output, since as much as 10 amperes can be flowing in the blower circuit. This reduced ammeter reading may be misinterpreted as a lack of sufficient output, even though the charging system is functioning properly. This discrepancy occurs because part of the alternator output is going through the blower motor circuit without passing through the test ammeter, since the two circuits are in parallel.

Notes

Figure No. 54

DIODE TESTS

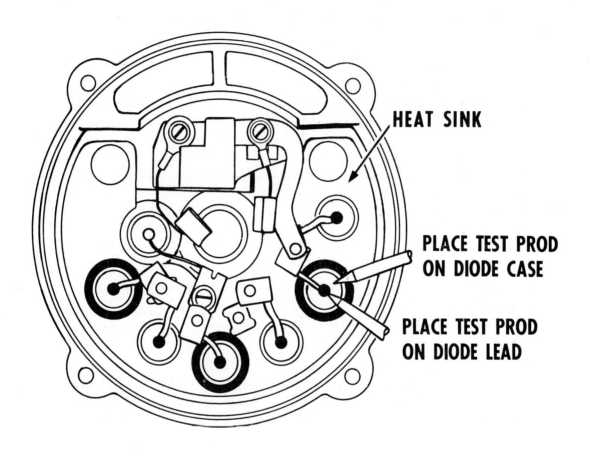

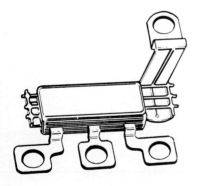
DIODE TRIO

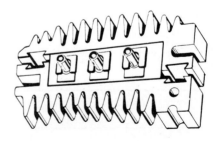
RECTIFIER BRIDGE

© Copyright 1986, Tune-Up Manufacturers Institute

DIODE TESTS

Due to the manner in which the three stator windings are connected, two windings are always being used at any one time. Only two diodes are being used at any one time. If any diode is defective, causing one phase of the three-phase winding to be missing, alternator output is decreased by approximately two-thirds, because any single missing phase undesirably influences both of the other phases.

The diodes on some makes of alternators can be tested with the alternator assembled and on the engine. Other makes have to be disassembled for diode testing.

Diodes can be readily checked using a diode tester. When using this test instrument, follow the manufacturer's instructions.

Diodes may also be tested with a 12-volt test lamp. Touch the prods of the test lamp leads to the diode case and diode lead, and then reverse the test lamp prods as previously explained. If the diode is good, the test lamp will light in only one test. If the lamp fails to light in both tests or lights in both tests, the diode is defective.

Another method of checking diodes is with an ohmmeter. After the stator leads have been disconnected, each diode can be tested for shorts and opens. Touch one ohmmeter prod to the diode case and the other prod to the diode lead, and observe the ohmmeter reading. Then reverse the ohmmeter prods and again observe the meter readings. A diode in good condition will have one high reading and one low reading. If both readings are very low or if both readings are very high, the diode is defective. Push and pull the diode lead *gently* while testing it to detect loose connections. Test all six diodes in the same manner. The readings of the diodes in the insulated heat sink will be opposite from those in the grounded heat sink or end frame

When a diode is found defective, it is advisable to replace all three diodes in the end frame or heat sink using the correct procedure and the proper removal and installation tools. Diodes must never be hammered into position as the impact can easily crack the silicon wafer. Be sure to test the condenser, as it may also be damaged.

Recently, the alternator rectifier bridge was introduced. The bridge contains all six diodes. This new design further simplifies the diode replacement operation. The bridge also contains the fins that are necessary for cooling the heat sink in which the diodes are mounted.

Figure No. 55

THE DIODE TRIO

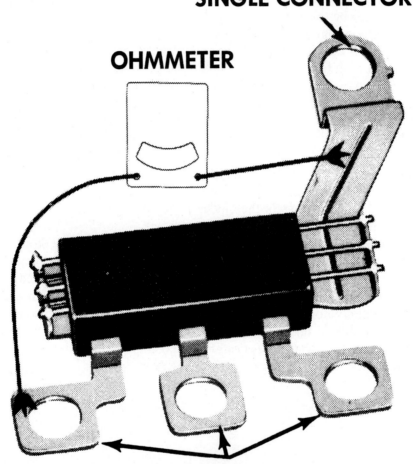

SINGLE CONNECTOR

OHMMETER

THREE CONNECTORS

TESTING A DIODE TRIO

DIODE TRIO

Some late-model alternators use, in addition to the regular diode assembly or rectifier bridge, a diode trio. It consists of three diodes contained in a single package, as shown in Figure No. 55, and is used to supply field current to the rotor. This unit is mounted internally in the alternator and must be removed for testing.

Testing is best accomplished with an ohmmeter, as shown in Figure No. 55. Connect one ohmmeter lead to the single connection on the end and then touch each of the three other connections. The readings should all be the same, infinitely high or very low. Then reverse the ohmmeter leads and repeat the test. These readings should be the opposite of the first group. If the ohmmeter gives the same reading with both connections, the diode trio is defective.

Figure No. 56

FIELD WINDING TESTS

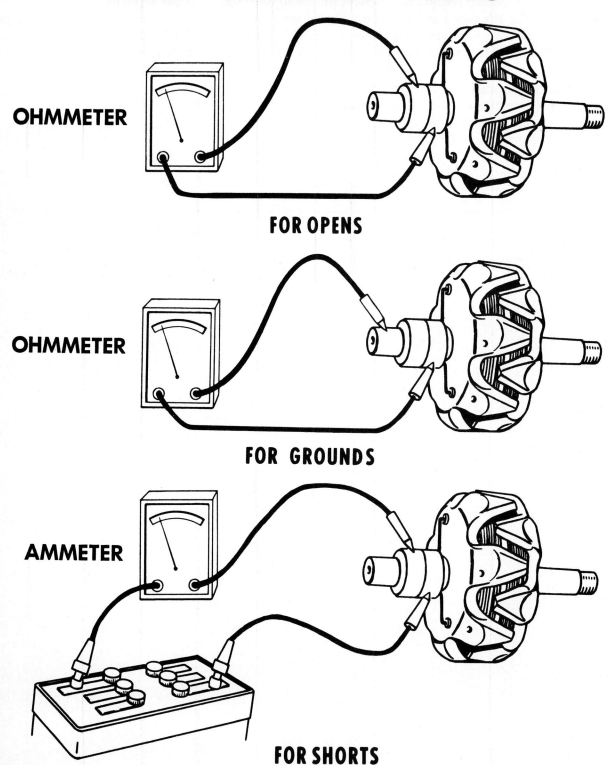

FIELD WINDING TESTS

The rotor field winding may be checked electrically for open circuit, ground circuit, and short circuit.

To check for open circuit, touch a 110-volt test lamp prod to each slip ring. If the lamp fails to light, the field winding is open.

To check for ground circuit, touch one 110-volt test lamp prod either slip ring and the other test lamp prod to the rotor shaft. If the lamp lights, the field winding is grounded.

To check for short circuit, connect a 12-volt battery and an ammeter in series with the two slip rings. The ammeter should indicate approximately 2 amperes. An ammeter reading above this value indicates a shorted field coil winding. High output alternators will have a higher field current draw.

If an ohmmeter is available, these three tests can be conducted by measuring resistance values.

Figure No. 57

STATOR WINDING TESTS

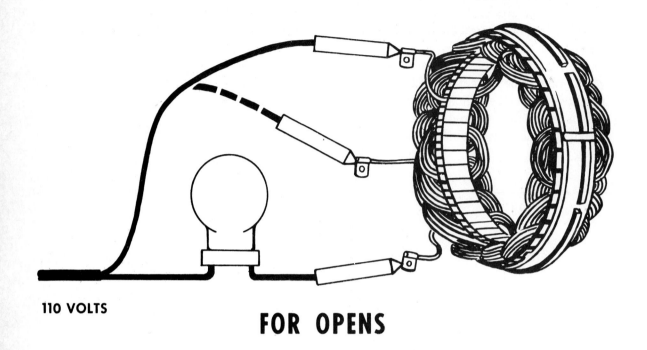

FOR OPENS

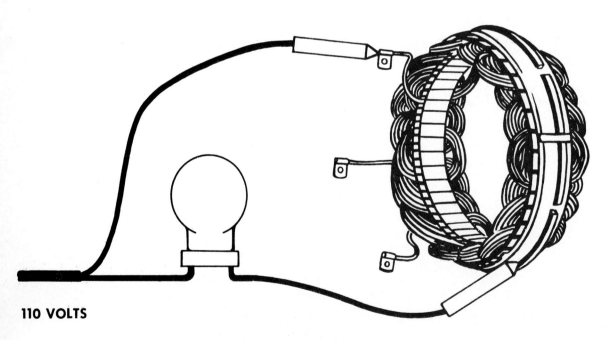

FOR GROUNDS

STATOR WINDING TESTS

Stator windings are checked for open circuit, ground circuit, and short circuit.

To check for open circuit, touch the prod of a 110-volt test lamp to the stator winding terminals as illustrated. If the lamp does not light, the stator winding is open. To complete the test, move one test lamp prod to the other stator winding terminal.

To check for a ground circuit, touch one prod of the 110-volt test lamp to the stator frame and the other test prod to any of the stator winding terminals. The test lamp should not light. If it does, the stator winding is grounded.

If an ohmmeter is available, these tests may be conducted by measuring resistance values.

A visual inspection for charred winding insulation should also be conducted at this time.

Figure No. 58

ALTERNATOR CHARGING SYSTEM SERVICE PRECAUTIONS

1. Always be *absolutely sure* that the battery ground polarity and the charging system polarity are the same, when installing a battery.

2. *Do not* polarize an alternator.

3. *Never* short across or ground any of the terminals on either the alternator or the regulator.

4. *Do not* operate an alternator on open circuit.

5. Booster battery *must be* correctly connected.

6. Battery charger *must be* correctly connected.

7. *Always* disconnect the battery ground cable before replacing or servicing electrical units.

ALTERNATOR CHARGING SYSTEM SERVICE PRECAUTIONS

Alternators are designed and constructed to give long periods of trouble-free service with minimum maintenance. To avoid accidental damage to the alternator, regulator, or charging system wiring, the following precautions should be observed:

1. *Always be **absolutely sure** that the battery ground polarity and the charging system polarity are the same, when installing a battery.* If a battery is hooked up backward, it is directly shorted across the alternator diodes. The high current flow can damage the diodes and even burn up the wiring harness. If battery post identification is not obvious, use a voltmeter across the posts to identify their polarity.

2. *Do **not** polarize an alternator.* The reason a dc generator is polarized is to excite the generator field to insure that the generator and battery will have the same polarity. Since the alternator develops voltage of both polarities, which the diodes automatically rectify, there is no need to polarize an alternator. In fact, damage to the alternator, regulator, or circuits may result from an attempt to polarize the alternator.

3. *Never short across or ground any of the terminals on either the alternator or the regulator.* Care should be exercised when working in the engine compartment to avoid accidental shorting of the alternator or regulator terminals. Shorting or grounding of the alternator or regulator terminals, either accidental or deliberate, can result in damage to the diodes, the regulator, and/or the wiring. Grounding of the alternator output terminal (Bat), even when the engine is not running, can result in damage since battery voltage is applied to this terminal at all times. Care should also be exercised when adjusting the voltage regulator to prevent accidental shorting.

4. *Do not operate an alternator on open circuit.* Operating the alternator while it is not connected to the battery or to any electrical load will cause the voltage developed to be extremely high. This high voltage can damage the diodes.

5. *Booster battery **must be** correctly connected.* When the booster battery is used to assist in engine starting, it must be connected to the car battery in proper polarity to prevent damage to the diodes. The positive cable from the booster battery must be connected to the car battery positive terminal, and the negative cable from the booster battery must be connected to a good ground (frame) of disabled vehicle. Positive to positive and negative to frame is the proper hook-up.

6. *Battery charger must be correctly connected.* Battery charger leads must be correctly connected to the battery, the positive charger cable to the positive battery post, and the negative charger cable to the negative battery post. Failure to observe this precaution will also result in damage to the diode rectifiers. When charging a battery, disconnect the battery cables before connecting the charger leads to the battery to prevent possible damage to the alternator. A fast battery charger should *never* be used as a booster for starting the engine in a car equipped with an alternator.

7. *Always disconnect the battery ground cable before replacing or servicing electrical units.* Disconnecting the battery ground cable is always advisable

when replacing electrical units or servicing electrical components. This precaution will prevent accidental shorting, which may result in damage to the diodes, regulator, or wiring. Remember, too, that if the battery ground cable is not disconnected, *make sure* the ignition switch is turned off when servicing the regulator since the alternator field circuit is connected to the battery through the ignition switch.

Notes

Figure No. 59

GENERATOR OPERATING PRINCIPLES

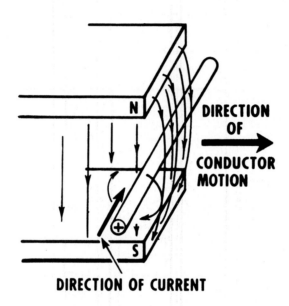

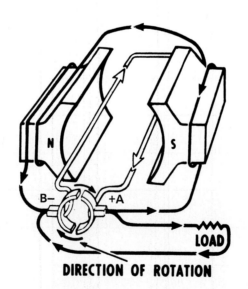

GENERATOR OPERATING PRINCIPLES

Before the development of solid-state devices, that is, diodes, the generator was the voltage source used in charging systems. Whenever the coil, called the *armature,* is rotated in the magnetic field flowing from the north to south poles, a voltage is generated in the windings. The magnetism flowing from the north to south pole is created by field coils, which are made of turns of copper wire wrapped around the north-south pole shoes.

All rotating electrical machines create ac. The battery charging system requires dc. The generator conversion from ac to dc was caused by a mechanical reversing switch called the *commutator.* As shown in Figure No. 59, the end of the coil marked A is connected to the positive brush. When this coil is rotated 180 degrees, point A will be negative, but through the use of the commutator it will be connected to the negative brush, thus giving a rectified dc output. For simplicity and purposes of illustration, the armature shown has only one coil and two commutator segments. Actually, armatures have many coils and a commutator segment connected to each coil end.

The amount of voltage developed will be determined by two variables, the speed of rotation of the armature and the strength of the magnetic field. The strength of the magnetic field is dependent on the amount of current flowing through the field coils. On a vehicle, the speed of rotation varies greatly, so voltage control is attained by varying the amount of current flowing through the field coils. The control of the field current is the function of the voltage regulator, and the generator voltage regulator works in the same manner as a regulator used with alternators; this was discussed under alternators.

Generators had a capacity to produce voltages that would create currents far larger than the generator was able to handle properly and, if not controlled, could result in generator burnout. To protect the generator, a device called a *current limiter* was used. This was a relay that would sense the generator output current and, as required, reduce the generator voltage to whatever value was necessary so that the output current would not exceed its safe rating.

Figure No. 60

GENERATOR TESTING

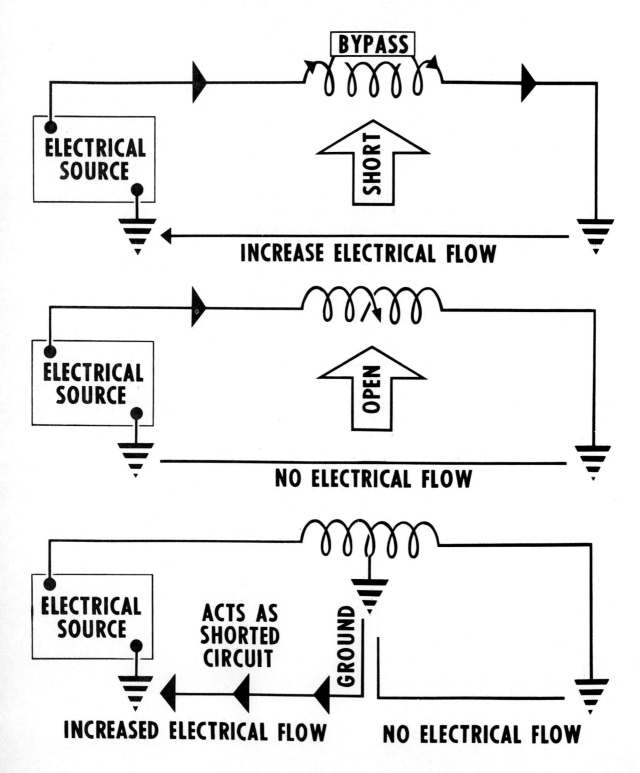

GENERATOR TESTING

The charging system, like any other system or component on the vehicle, requires periodic service or maintenance to assure top operating efficiency. Should trouble occur in this system, the cause and location of this trouble can readily be determined through a systematic test procedure. To efficiently test charging system components so that proper service operations may be performed, it is necessary to have a clear understanding of electrical troubles. Basic malfunctions can be classified into four groups. *Short* circuits, *open* circuits, *grounded* circuits, and circuits with abnormally *high* or *low resistance*.

A short circuit is any accidental contact that permits the current to bypass a portion of the electrical circuit. This condition is present when one or more windings on a coil are bypassed due to insulation failure. A short circuit results in a lower-than-normal circuit resistance, thereby permitting a higher-than-normal current flow.

An open circuit is an undesired break in the circuit. A break can occur in any one of a number of locations in the circuit, such as coil windings, wires, or connections, resulting in an inoperative circuit. No current will flow through an open circuit.

A grounded circuit is an undesired connection that bypasses part or all of the electrical units, from the insulated side to the ground side of the circuit. In a lighting circuit, for example, should a ground occur between the battery and the lamps, the load on the battery would become unreasonably high while the lamps would fail to light. In general, a grounded circuit results in a higher-than-normal current flow produced by a proportionate reduction in circuit resistance.

A circuit with abnormally high resistance is one containing resistance of a nature that increases the total resistance of the circuit. Poor or loose connections, corroded connections, and frayed or damaged wires are examples of conditions causing high resistance. Should this condition exist, current flow will be less than normal because of the increase in circuit resistance.

Notes

6

THE IGNITION SYSTEM

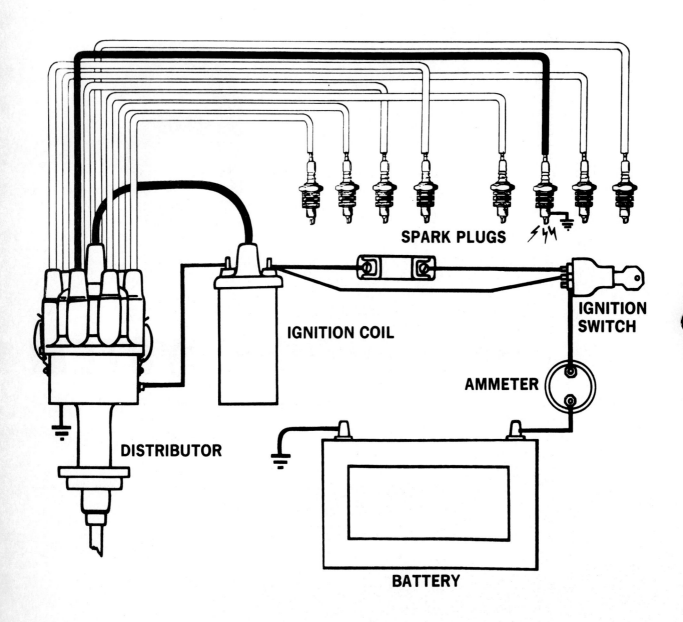

Figure No. 61

THE IGNITION CIRCUIT

IGNITION CIRCUIT (BREAKER-POINT TYPE)

The internal combustion engine operates through the forces created by expanding gases in the combustion chambers. These gases are the product of the burning air-fuel mixture, which was ignited by a high-voltage spark.

The function of the ignition system is to produce a high-voltage surge, and to deliver it to the proper spark plug at the correct instant to ignite the air-fuel mixture compressed in the cylinder.

The ignition system consists of the following components:

Battery	Distributor assembly	
Ignition switch	Body	Breaker points
Ballast resistor	Cap	Condenser
or ballast wire	Rotor	Advance mechanisms
Ignition coil	High-tension leads	
Spark plugs		

The battery is the source of power which supplies low voltage to produce a current flow in the ignition primary circuit.

The ignition switch is simply an on-off switch to complete the ignition circuit between the battery and coil. When the ignition switch is closed, current will flow through the coil-distributor (primary) circuit and return by way of the car frame or engine block to the battery. The ignition switch also serves as a bypass switch during engine starting.

The ballast resistor in the ignition primary circuit is designed to permit the proper amount of current flow for all driving conditions. During cranking, however, it is bypassed to permit full battery voltage and maximum current flow through the coil for quick starting.

The ignition coil transforms or *steps up* the low battery voltage to a voltage high enough to jump a spark gap at the spark plug.

The distributor interrupts the flow of current in the primary winding by the action of the breaker points. It also distributes the high-tension current developed by the coil through the rotor and the distributor cap to the proper spark plug at the correct instant.

The high-tension or secondary leads conduct the high voltage produced by the ignition coil to the distributor, and from the distributor to the spark plugs.

The spark plugs provide a spark gap in the combustion chamber. When high voltage jumps across the gap, the air-fuel mixture is ignited.

© Copyright 1986, Tune-Up Manufacturers Institute

Figure No. 62

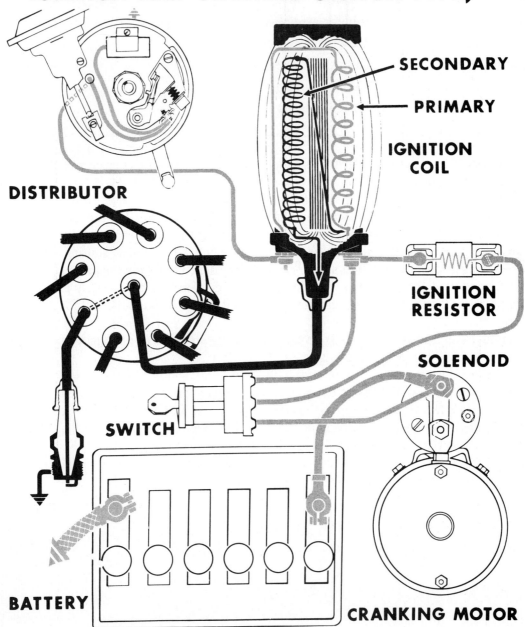

IGNITION SYSTEM

The ignition primary circuit is also called the low-voltage or the low-tension circuit. This is the circuit through which current flows at battery or generator voltage. The ignition switch, resistor, ignition coil primary winding, breaker points, and condenser are all in the primary circuit.

The secondary ignition circuit is also called the high-voltage or high-tension circuit. The secondary coil winding, distributor rotor, distributor cap, high-tension leads, and the spark plugs are all part of the secondary circuit.

When the ignition switch is turned on and the breaker points are closed, current flows in the primary circuit. As current flows through the primary winding of the ignition coil, a strong magnetic field is produced in the coil. When the breaker points open, current through the primary winding of the coil is stopped, and the magnetic field around the coil winding collapses. These collapsing lines of force cut across both the primary and secondary windings, inducing a very high voltage in the secondary winding. The high voltage so induced forces current to jump the spark-plug gap.

The primary ballast resistor is essentially a current-compensating device consisting of a resistor unit or wire located in the primary ignition circuit. The compensating action is obtained because, at low engine speeds, the current flows for longer periods of time. This heats up the resistor, thereby raising its resistance and reducing current flow. This action serves to keep the coil primary winding cooler and improves distributor breaker point life. At high speeds, the current flows for shorter periods of time, which lets the resistor cool and increases the current flow in the primary winding of the coil. This action permits maximum secondary voltage to be obtained.

Because of the lowered battery voltage resulting from the starter load on some vehicles, the ballast resistor is bypassed while the starting system is in operation. This is done to provide higher secondary voltage for starting.

Figure No. 63

DISTRIBUTOR ASSEMBLY

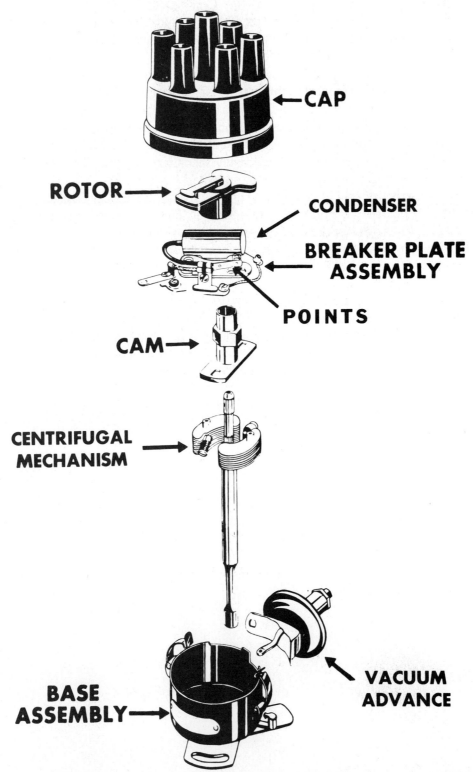

DISTRIBUTOR ASSEMBLY

The distributor assembly is made up of several subassemblies: the cap, rotor, breaker plate assembly supporting the breaker points and condenser, cam, centrifugal mechanism, vacuum advance unit, and the distributor base or body.

The distributor cap and rotor are used to distribute the high-voltage current developed in the coil to each spark plug in firing-order sequence. The distributor cap is keyed to the distributor body to maintain the correct relationship between the distributor cap towers and the rotor. The rotor is keyed to the distributor cam to be in the correct position to transfer the high-voltage current to one of the distributor cap high-voltage towers when the points open.

The breaker plate supports the breaker points at the correct position for the cam to open them. In most distributors, the breaker plate is movable and is mounted on a center bearing or side pivot. Since a moving part can develop high resistance at the bearing surface, a flexible insulated wire is used to connect the points to the primary circuit. Also, a grounding wire may be used between the plate and distributor housing.

The distributor cam opens the breaker points to interrupt the primary circuit. The cam has the same number of lobes as the engine has cylinders, as the points must be opened to produce a high-voltage surge for each power stroke.

The centrifugal advance mechanism varies the position of the cam in relation to engine speed. The mechanism has weights that are thrown out against calibrated springs to advance the breaker cam as speed increases. The advance mechanism is usually mounted in the distributor housing below the breaker plate. Some late-model distributors have the centrifugal advance mechanism above the breaker plate and below the rotor.

The vacuum advance unit advances the ignition timing in addition to the advance provided by the centrifugal unit. The vacuum unit is controlled by manifold vacuum and consists of a metal chamber in which a flexible airtight diaphragm is located. A link extends from one side of the diaphragm to the breaker plate assembly or distributor housing. If the distributor is equipped with a movable breaker plate, the vacuum advance mechanism rotates the plate to advance the ignition timing. If the plate is rigidly mounted in the distributor, the vacuum advance unit rotates the complete distributor assembly for advance.

Figure No. 64

DWELL ANGLE

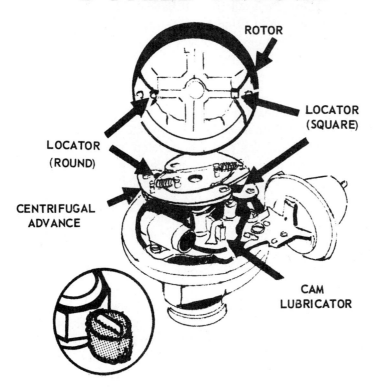

CONTACT POINT DWELL TEST

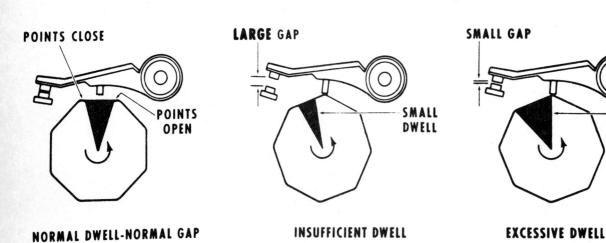

NORMAL DWELL-NORMAL GAP   INSUFFICIENT DWELL   EXCESSIVE DWELL

DWELL ANGLE

Dwell angle is the number of degrees the distributor cam turns during which time the breaker points are closed. During the dwell period, a magnetic field is built up in the primary winding of the coil. However, time is required to build up a full-strength magnetic field. When a full-strength magnetic field is produced, the coil is said to be *saturated*. To be assured of coil saturation at low engine speeds presents no problem because of the relatively slow rotation of the distributor cam. At high speeds, however, unless the distributor points are adjusted to provide a sufficient dwell period, coil saturation will not be attained.

In a six-cylinder engine running at idle speed of 400 rpm, the ignition system must produce 20 sparks per second to fire all the cylinders. With an engine operating at this speed, dwell angle is not very critical because more-than-sufficient coil saturation time is available. However, with the engine running at 4000 rpm, or the equivalent of 90 miles per hour or more, 200 sparks per second would be required to fire all the cylinders. This is the speed at which dwell becomes extremely critical. If the dwell was reduced only slightly from the required amount, the engine would begin to misfire at high speed because the coil does not have time to become sufficiently saturated.

Although dwell is not critical at low speed, point gap becomes very important. With the engine cranking, there must be sufficient point gap or the points will arc excessively and the engine will not start readily. Also, if an engine is operated with too little point gap at low speed, the points will deteriorate rapidly. If the points open slowly, and do not open wide enough, an arc will continue across the contact points using energy that would normally create a spark at one of the spark plugs. When an arc occurs, the engine usually misfires because the energy of the primary circuit is dissipated, preventing sufficient secondary circuit voltage buildup.

Dwell angle adjustment directly affects ignition timing. Under certain conditions the rubbing block on the movable breaker arm may wear. As a result, the dwell angle increases, which in turn causes the ignition timing to be late.

One of the largest single causes of breaker-point failure is the lack of cam lubricant. Point rubbing block wear can be appreciably reduced by applying a thin film of high-temperature cam lubricant to the distributor cam when servicing the distributor. It is important that the proper lubricant be used, since it must be able to adhere to the cam surface at high cam speed, resist melting at high temperature, resist chemical reaction with the polished steel cam, effectively control moisture to prevent rust formation on the cam, and resist drying out with age.

Some distributors are fitted with a cam lubricator, the wick of which is impregnated with a special lubricant. At specified intervals the cam lubricator should be rotated 180 degrees (or end for end) with the lubricator just touching the cam lobes. At every other service interval, the lubricator should be replaced. Lubricant should never be added to the lubricator or used on the cam of the distributor equipped with this device.

Ignition point spring tension plays an important part in the performance of the ignition system, and must be within specified limits. Excessive pressure causes rapid rubbing block and cam wear, while insufficient pressure will permit high-speed point bounce, which, in turn, will cause arcing and burning of the points and misfiring of the engine.

Side-Pivoted Breaker Plate

Before testing and adjusting the dwell angle on distributors that have a side-pivoted breaker plate, as on Chrysler, Ford, Autolite, and Prestolite distributors, it is recommended the distributor vacuum line(s) be disconnected and plugged.

On some 1970 to 1972 Chrysler-built eight-cylinder engines equipped with a distributor solenoid, the solenoid lead should also be disconnected at the connector when testing the dwell angle. The reason for this specific procedure is that the dwell-angle specification is based on "No vacuum advance" and "No solenoid retard." Therefore, the vacuum line and the distributor solenoid must be disconnected. *Do not* attempt to disconnect the lead at the solenoid.

The above procedures are required because the distributor shaft and the cam rotate about the center point of the shaft. The breaker points, however, are mounted on the breaker plate, which pivots about its anchor. The points and the cam are therefore moving about two different centers or pivots. If the dwell angle is tested or adjusted while the vacuum lines are connected, the breaker plate is rotated for spark advance, causing the breaker gap to increase. This action results in a decreased dwell angle because the points are closed for a shorter time.

This is also the reason why the permissible dwell variation on this type of distributor is approximately double that allowed for the center-pivoted plate type.

To accurately test and set the dwell angle on this type of distributor, it is important that the recommended procedure be followed.

Remember—the selection of a quality, precision-manufactured set of breaker points, their proper installation, and precise adjustment is the most critical single factor in the efficient operation of the ignition system.

Notes

AS ENGINE SPEED INCREASES SPARK MUST BE TIMED EARLIER

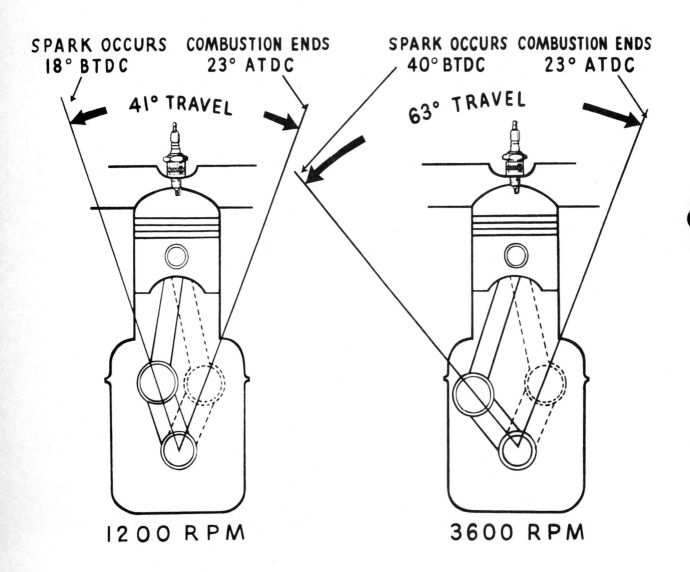

SPARK ADVANCE TIMING

To obtain full power from combustion, the maximum pressure must be reached just as the piston passes top dead center, and combustion must be completed by approximately 23 degrees after top dead center (ATDC).

The fuel mixture ignited in the combustion chamber does not explode. It burns rapidly until the fuel is consumed. The time required for complete combustion is a small fraction of a second. For this reason, ignition must take place before the piston passes top dead center.

As engine speed increases, the piston moves through the compression stroke more rapidly, but the burning rate of the fuel mixture remains virtually the same. To compensate for the higher piston speed, ignition must occur earlier in the compression stroke.

During the combustion process at 1200 rpm, the crankshaft travels through 41 degress of rotation, from the point of ignition to 23 degrees after top dead center. The spark must occur at 18 degrees before top dead center (BTDC). The same engine running at 3600 rpm will require 63 degrees of crankshaft rotation to complete combustion by 23 degrees after top dead center. This would require that the spark occur 40 degrees before top dead center. This is the reason why spark advance is such an important factor in efficient engine operation.

Setting the basic spark timing will be covered in detail in the tune-up procedure section of this course.

Figure No. 66

CENTRIFUGAL ADVANCE MECHANISM

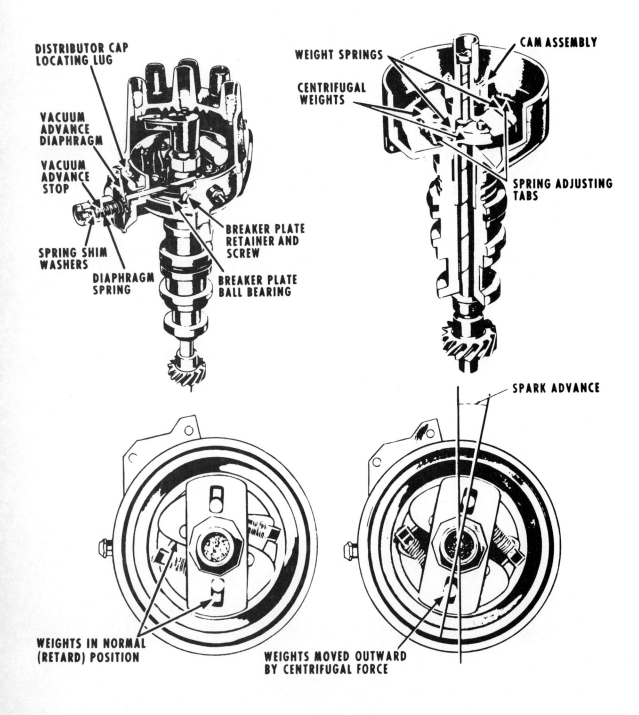

CENTRIFUGAL ADVANCE MECHANISM

Most distributors are equipped with a centrifugal-type advance mechanism. This mechanism automatically advances spark timing in correct relation to any engine speed above idle. The distributor cam is linked to the distributor shaft through the advance mechanism in such a manner that, as distributor speed increases, the weights move out due to centrifugal force. This causes the cam to rotate several degrees ahead of the shaft, opening the breaker points earlier in the compression stroke and thereby advancing the ignition timing. Springs return the weights to their retarded position as engine speed is decreased. This type of advance mechanism is actuated only by engine speed.

The amount of advance required at certain speeds will vary depending upon engine design, compression ratio, air-fuel ratio, and the octane rating of the fuel. Individual advance curves for different engines are designed into the contour of the cam and its slots and by using weight return springs of various lengths and tensions.

A quick method of checking if the centrifugal advance mechanism is working is to twist the rotor in the direction of rotation and then quickly release it. The rotor should snap back into its released position. Failure to return to its released position indicates broken return springs. A slow sluggish return action indicates a gummed or rusted condition of the advance mechanism. Either condition must be corrected before a tune-up can be effective.

Another method of checking the action of the centrifugal mechanism is to disconnect the vacuum line and tape its opening, and slowly accelerate the engine while observing the timing mark position with a power timing light. If the mechanism is functioning, the timing mark will move against engine rotation as engine speed is increased.

Figure No. 67

SPARK ADVANCE

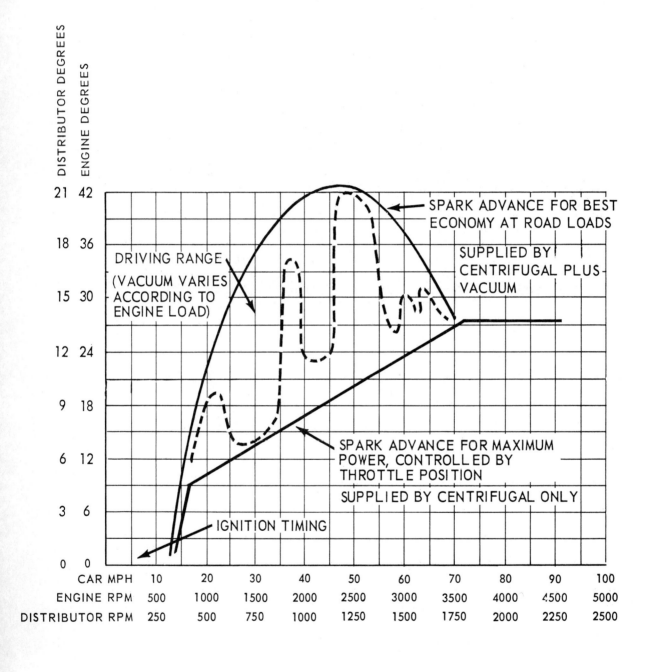

SPARK ADVANCE

The centrifugal advance mechanism adjusts spark timing according to changes of engine speed, while the vacuum advance unit adjusts spark timing in relation to engine load. The two devices operate independently, but the total spark advance is dependent on both.

An engine operating at 1500 rpm may be running under very light load with the throttle open only a small amount. This would be true if the car were being driven at a fixed speed on a level road. On the other hand, at the same speed, the engine may be under full load, as would be true if the car were being driven up a steep hill with the throttle wide open. In the first example, engine speed at 1500 rpm (engine under light load), the manifold vacuum would be high. This would operate the vacuum advance, and the total spark advance would be the sum of the centrifugal advance plus the vacuum advance. In the second example, at the same engine speed but operating under full load, very little vacuum exists in the manifold. Because very little vacuum is available, the vacuum advance unit would return to its retarded position. Therefore, the total spark advance would depend on the operation of the centrifugal advance unit only.

Under any operating condition, the total spark advance is dependent upon the speed of the engine and on the load under which the engine operates. The higher the speed, the greater the advance, and also, the lighter the load, the greater the advance. At any speed, the load on the engine may vary from zero to full load depending on the position of the throttle plate. When the engine is operating at high speed, the throttle will be open, which results in low manifold vacuum. Spark advance is controlled entirely by the centrifugal advance mechanism at this time.

The total spark advance at any engine speed is the initial advance, plus the centrifugal advance, plus the vacuum advance.

Figure No. 68

IGNITION COIL CONSTRUCTION

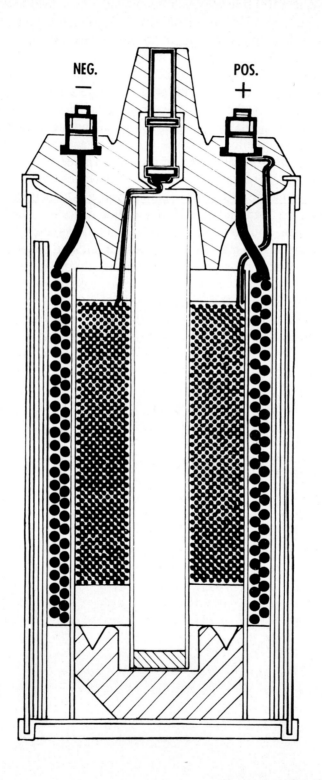

IGNITION COIL

Construction

An ignition coil is composed of a core, two windings, a housing, and a mounting bracket. The core of the coil usually consists of thin, soft iron strips or laminations. Their purpose is to increase the efficiency and output of the coil by promoting faster and more complete coil magnetic saturation. The soft iron core readily conducts magnetic lines of force, so less energy is used than if the magnetic lines of force had to travel through an air core.

The two windings are identified as a *primary* winding and a *secondary* winding. The primary winding consists of approximately 250 turns of relatively heavy wire, which is insulated with a special varnish. The secondary winding is wound inside the primary winding and consists of approximately 20,000 turns of very fine varnished wire. The many layers of the secondary windings are insulated from each other by high-dielectric paper. One end of the secondary winding is connected to the high-tension tower; the other end is connected to one of the primary terminals inside the coil.

Ignition coils are often filled with oil or special compound to provide additional insulation and to help dissipate the heat that is created by the transformation of battery voltage. The dissipation of heat is very important in an ignition coil, as heat tends to weaken insulation. An insulation breakdown results in partial or total coil failure.

Ignition Coil Action

When the ignition switch is turned on and the electronic switch or breaker points are closed, current flows in the primary circuit. As current flows through the primary winding of the ignition coil, a strong magnetic field is produced with the aid of the core. When the breaker points open, current ceases to flow through the primary windings of the coil and causes the magnetic field to collapse across the many thousands of turns of wire in the secondary winding. This action induces a very high voltage in the secondary circuit, which forces current to jump the rotor and spark-plug gaps.

When a piece of wire is connected across a source of voltage, current will immediately reach a maximum value determined by the resistance of the wire itself. But this is not true when the same wire is wound into a coil as in the primary winding of the ignition coil. This characteristic of a coiled conductor is called *reactance* or *counter-electromotive force,* and is due to the self-induced voltage in the coil.

When the electronic switch or breaker points close, current starts to flow in the primary winding. As the magnetic field begins to build up, the lines of force cut through the primary winding and induce a voltage that opposes battery voltage. Therefore it takes a definite period of time for the primary current to reach a maximum rate of flow after the electronic switch or breaker points close. This period of time is called a *buildup time.* When maximum current is flowing in the coil winding, the maximum magnetic field is present, and the coil is said to be fully *saturated.*

If the breaker points in a normal ignition system remain closed for too short a period of time, maximum current flow will not be reached in the primary circuit, and the maximum magnetic strength will not be attained. As a result, when the breaker points open, there will be less lines of force to cut through the secondary winding, and coil output voltage will be reduced. This can cause the engine to misfire under certain operating conditions.

Reactance or counterelectromotive force not only opposes the buildup of current through the primary circuit, but also opposes any attempt to stop the flow of current. As the breaker points open, the magnetic field starts to collapse. The lines of force cut through the primary winding, but in the opposite direction from the buildup. This causes an induced voltage in the primary winding, which is in the same direction as battery current and tends to keep current flowing. If current continues to flow when the breaker points open, there will be an arc between the breaker points. This arc has two very detrimental effects. First, it causes a transfer of metal from one point to the other, resulting in point pitting. Second, unless the flow of primary current is stopped quickly, the magnetic field will collapse gradually, and the secondary winding output voltage will be considerably reduced.

To control the arc that takes place between the points as they separate and to quickly stop the flow of primary current to develop maximum coil voltage, a condenser is connected across the points.

Ignition Coil Replacement

As previously stated, the ratio of coil secondary turns to primary turns is approximately 100 to 1. Therefore, a typical coil for a normal ignition system would have 200 turns of primary winding and 20,000 to 26,000 turns of secondary winding.

The coils used in transistorized ignition systems have a turns ratio of either 275 to 1 or 400 to 1. Because of the high current-carrying capacity of the heavier-gauge transistor coil primary winding, approximately 95 turns of primary winding will be used in a coil having 26,000 turns of secondary winding. These coils use an epoxy compound to provide adequate insulation.

The coils used in certain high-energy ignition systems also have a turns ratio of approximately 100 to 1. These coils are designed to be used with a high-energy distributor cap and rotor.

It is very important that when coil replacement is required the proper replacement coil be installed on the engine. Mixing normal and transistor system coils or coils of the wrong polarity will be quickly reflected in poor engine performance or total ignition failure.

Notes

Figure No. 69

COIL POLARITY

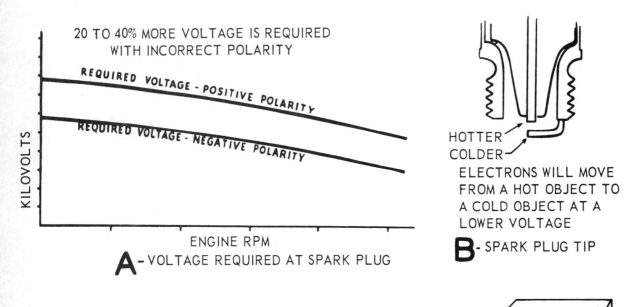

A - VOLTAGE REQUIRED AT SPARK PLUG

20 TO 40% MORE VOLTAGE IS REQUIRED WITH INCORRECT POLARITY

B - SPARK PLUG TIP

ELECTRONS WILL MOVE FROM A HOT OBJECT TO A COLD OBJECT AT A LOWER VOLTAGE

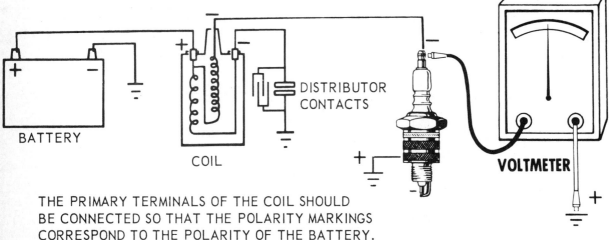

C - CORRECT COIL CONNECTIONS

THE PRIMARY TERMINALS OF THE COIL SHOULD BE CONNECTED SO THAT THE POLARITY MARKINGS CORRESPOND TO THE POLARITY OF THE BATTERY.

CORRECT SPARK PLUG POLARITY WILL RESULT IF THE COIL IS PROPERLY CONNECTED.

GROUND THE VOLTMETER POSITIVE LEAD AND TOUCH THE NEGATIVE LEAD TO THE SPARK PLUG TERMINAL.

COIL POLARITY

To keep the required firing voltage of the ignition system as low as possible, the ignition coil must be connected for the correct polarity. The primary terminals of the coil should be connected so that the polarity markings correspond to the polarity of the battery, with the distributor connection considered as the ground. This will cause current flow through the spark plug from the center electrode to the ground electrode. This spark polarity, or secondary polarity, requires a lower voltage to fire the spark plugs since the electrons will be emitted from the hotter center electrode more easily than from the cooler ground electrode.

If secondary polarity is reversed, 20 to 40 percent more voltage is required to complete the secondary ignition circuit.

The polarity of the ignition coil can be checked with the aid of a voltmeter. Set the voltage range switch to the highest scale, connect the positive voltmeter clip to a good ground, and touch the negative voltmeter lead to any spark plug terminal. The meter needle should move upscale. You are not interested in a meter reading, only in the direction the needle moves. Should the needle move to the left or downscale, reverse polarity is indicated. This means the leads to the coil primary terminals are reversed, the battery is installed backward, or the wrong coil is being used.

Ignition coils are not wound as negative or positive coils. Since *all* spark plugs have a positive ground, regardless of whether the vehicle's electrical system is negative ground or positive ground, it is advisable to make the spark-plug polarity test every time service is performed on the ignition system or whenever the battery cables have been disconnected.

Figure No. 70

COMPARISON OF AVAILABLE AND REQUIRED SECONDARY VOLTAGE

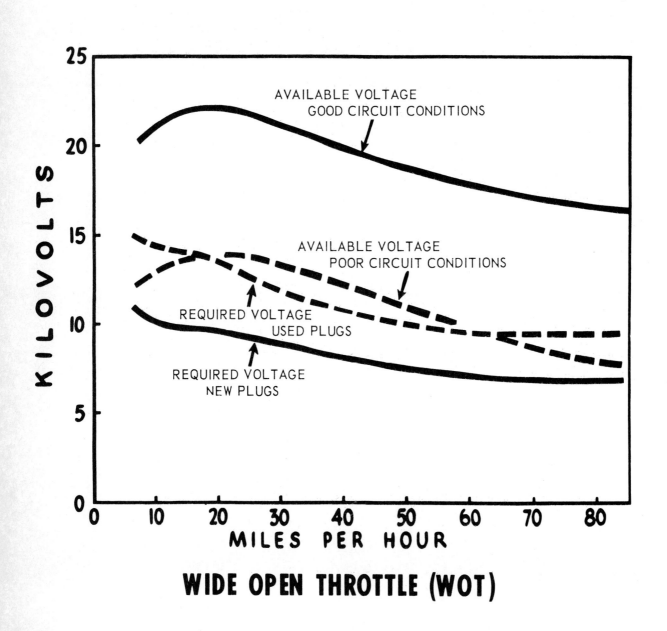

AVAILABLE VOLTAGE

Efficient ignition system performance requires that the available voltage, indicated by the upper solid line in Figure No. 70, be always higher than the required voltage, indicated by the lower solid line. If at any engine speed or load the required voltage exceeds the available voltage, ignition failure results. This condition is indicated by the crossing of the dotted lines.

Any condition that reduces the safety margin between the available and required voltage must be corrected before a tune-up can be considered successful. Reverse coil polarity, fouled spark plugs, plugs with eroded gaps, pitted breaker points, misaligned breaker points, improperly adjusted breaker points, leaking condenser, eroded distributor cap terminals, corroded distributor cap towers, defective rotor, loose spark-plug cable terminals, damaged resistor cables, or loose primary lead connections are all conditions that either introduce resistance into the ignition system, thereby reducing the system's available voltage, or increase the required voltage beyond the system's output capabilities.

Figure No. 71

CONDENSER CONSTRUCTION

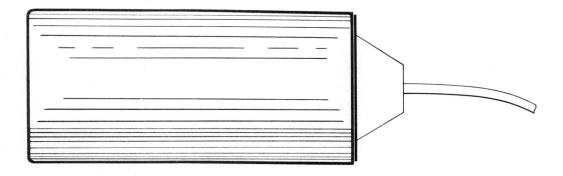

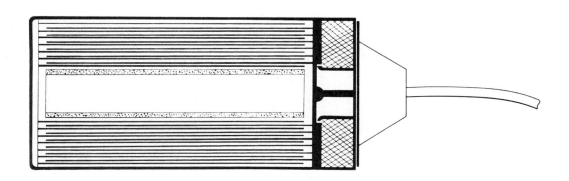

CONDENSER CONSTRUCTION

The condenser is constructed of layers of aluminum foil, insulated from each other by layers of high-dielectric insulating material. One layer of aluminum foil extends beyond the insulating material on one side; the other layer of foil extends beyond the insulating material on the other side. The layers of aluminum foil and insulating material are then rolled into a tight cylinder and inserted into the condenser case. The layer of aluminum foil extending at one side will contact the bottom of the case and represents the ground terminal of the condenser. The other layer of foil will contact a disc, which is connected to the insulated lead of the condenser.

Current does not flow through a good condenser. If it does, the two layers of foil are touching or there is a hole in the insulating material. If either condition exists, the condenser is defective and must be replaced.

CONDENSER ACTION

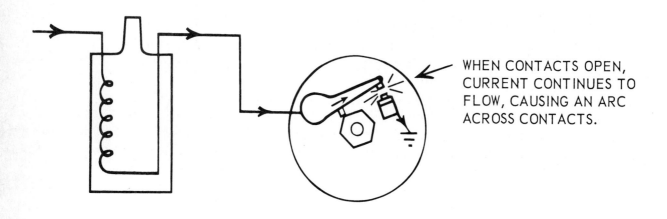

WHEN CONTACTS OPEN, CURRENT CONTINUES TO FLOW, CAUSING AN ARC ACROSS CONTACTS.

NO CONDENSER IN PRIMARY CIRCUIT

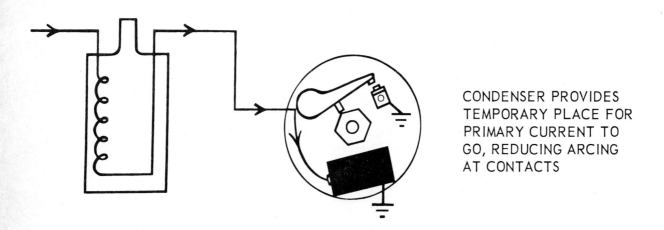

CONDENSER PROVIDES TEMPORARY PLACE FOR PRIMARY CURRENT TO GO, REDUCING ARCING AT CONTACTS

WITH CONDENSER IN PRIMARY CIRCUIT

CONDENSER ACTION

The functions of the ignition condenser are to reduce the amount of arcing across the points, thereby preventing excessive metal transfer from one point to the other, and to quickly stop the flow of current in the primary coil winding so that maximum voltage can be generated in the secondary coil winding.

The condenser can store a certain amount of electrical energy. When the breaker points open and the induced voltage in the coil tries to keep current flowing across the breaker points, the condenser will absorb the electrical energy until the breaker points have opened sufficiently so that an arc cannot occur. By the time the condenser becomes fully charged, the points have opened too far for arcing to take place.

By preventing the arc from occurring across the breaker points, the condenser brings the primary current flow to a sudden stop. This causes a very sudden collapse of the magnetic field, so that the lines of force cut through the windings in the coil with great speed. The voltage induced in the secondary winding forces current to jump the spark-plug gap. The charge in the condenser surges back in a reverse direction through the primary circuit across the battery and builds up on the opposite plate of the condenser. The condenser then discharges in the opposite direction and charges the condenser once more in the original direction. Each time the condenser discharges part of the energy is lost in overcoming the resistance of the circuit so that the oscillating current will die out, or nearly so, before the contact points close for the next buildup.

Condenser Testing and Selection

Condensers are tested for resistance, capacity, and leakage. The resistance test reveals the presence of any loose or high-resistance connections in the pigtail or the case. The capacity test checks the microfarad capacity of the condenser. The insulation test stresses the insulating material with about 500 volts while the tester meter indicates the presence of current leakage through the insulation. A condenser that tests defective in any test must be replaced.

The testing of a used condenser will reveal the presence of any defects, but it cannot measure the amount of useful life left in the condenser. For this reason it is advisable to always install a new condenser when replacing the breaker points.

Some transfer of metal from one breaker point to the other is a normal action that occurs in all ignition systems. The function of a condenser of correct capacity is to prevent any excessive metal transfer. The capacity of the condenser is an important specification. It has been selected to match the particular requirements of a given ignition system. When replacing the condenser, be sure the new condenser has the microfarad (mF) capacity recommended in the specifications. Typical specifications are 0.18 to 0.23 mF; 0.21 to 0.25 mF; 0.25 to 0.285 mF.

When a good condenser of the correct capacity is employed and a condition of excessive point metal transfer nevertheless exists, the condition may be due to the following:

1. Excessive primary or secondary circuit resistance in the ignition system, which will upset the system balance.

2. A high voltage regulator setting, which causes increased primary current flow, resulting in overheating and burning of the breaker points.
3. Incorrect breaker point dwell angle.
4. Continual high-speed operation.
5. Frequent periods of prolonged idling.
6. Grease or oxidation on housing of condenser making for poor ground contacts.

Locating and servicing the cause of the trouble will prevent a recurrence of the condition. In the case of the last two conditions, more frequent tune-ups should be recommended.

Notes

Figure No. 73

DISTRIBUTOR CAP AND ROTOR CONSTRUCTION

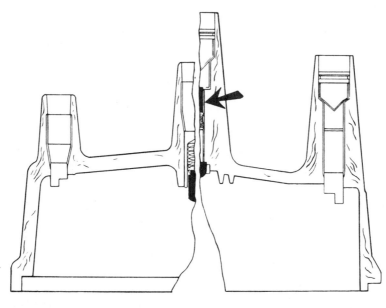

REGULAR CAP **RESISTOR CAP**

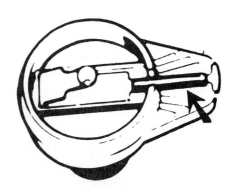

REGULAR ROTORS **RESISTOR ROTOR**

© Copyright 1986, Tune-Up Manufacturers Institute

DISTRIBUTOR CAP AND ROTOR

Distributor Caps

A distributor cap is constructed of an insulating and high-temperature dielectric material with metal inserts that are cast into the cap to receive the secondary wire from the coil to the distributor cap, and to receive the spark-plug wires. These metal inserts extend downward inside the distributor cap so that the distributor rotor can provide a path between the center terminal and an outside terminal of the distributor cap.

A resistor-type cap containing built-in suppression usually includes a carbon resistor integral with the terminal in the cap coil tower. Due to the change characteristics of the carbon resistance over a period of time, it is advisable to change the distributor cap and rotor when performing a breaker-point replacement.

Distributor Rotors

Distributor rotors are usually constructed of materials similar to those used in distributor caps. A metal strip is used to form a conductor which contacts the center button of the distributor cap and carries the high-tension current to the proximity of one of the secondary wire terminal posts inside the cap. This metal strip does not actually close the circuit between the center button and the outer terminal. There is always an air gap between the end of the rotor tip and the distributor-cap terminals. This gap is very small so that not more than 2000 to 3000 volts is required to carry secondary ignition current across this gap.

Some rotors incorporate a carbon resistor in their construction to provide ignition suppression. Due to the fact that carbon resistors occasionally undergo a change, unwanted resistance may be introduced by the loss of proper contact at the ends of the resistor. It is advisable, therefore, to change both the rotor and the distributor cap when replacing breaker points. Be sure to replace resistor-type rotors and caps with similar-type units to retain the balance of the ignition system.

Rotors are keyed to the distributor shaft, which places them correctly with relationship to the distributor cam. The relationship between the rotor and the distributor cam is critical, in that the rotor tip must be passing one of the secondary ignition wire terminals inside the cap at the time the ignition points open. To properly position the rotor on the centrifugal weight assembly of the General Motors V-8 engines and to prevent rotor breakage, be sure to observe the shape of the locators on the underside of the rotor—one is round and one is square. They fit into similarly shaped receptacles in the weight base for proper rotor position.

Figure No. 74

SECONDARY CIRCUIT SUPPRESSION

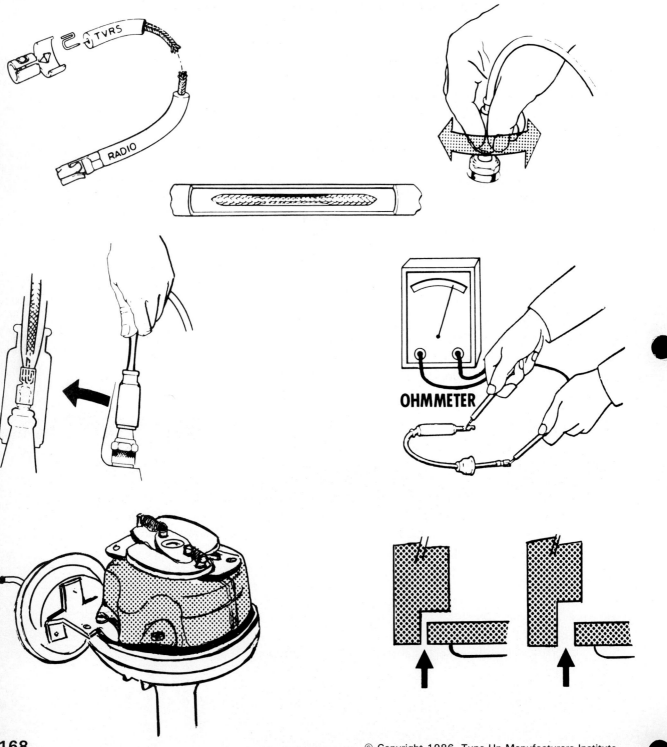

SECONDARY CIRCUIT SUPPRESSION

Any circuit which includes an electric arc or a spark will radiate electrical noise into the surrounding space unless some form of noise suppression is included in the circuit. The secondary circuit of an automotive ignition system, with its multiple spark gaps, can cause very serious interference with radio and television reception if it does not include suitable noise-suppression devices. In most vehicles, resistance or impedance (inductance) is added to the secondary circuits which couple the ignition coil to the spark plug through the distributor. This will prevent the flow of the high-frequency currents which cause radio-frequency interference (RFI). The most common suppression method is to put the resistance or impedance (inductance) in the secondary wires themselves.

Note: Ignition wires should never be "yanked" from the spark plugs. Instead a boot puller tool should be used to remove the spark-plug boot from the spark plug. If no tool is available, the spark-plug boot should be gently twisted to break the seal between the boot and the spark plug insulator. The wire should then be gently lifted from the plug.

Separate resistors or inductors also can be mounted at the spark-plug end of the secondary wires, in the distributor cap, or in the stem of the spark plug, as a means for reducing interference. While the addition of a few thousand ohms in the secondary wires of the ignition system will not adversely affect engine performance, excessive resistance can reduce spark intensity due to limiting of spark current. This can cause misfire and poor performance, particularly with some engines using a high air to fuel ratio to reduce emissions. Under ideal circumstances, resistance added to spark-plug wires would only limit current through the spark plug after it fires and not reduce the prefiring voltage. Under field conditions, there is always some electrical leakage to ground (chassis) due to contamination, fouled plugs, and deterioration of insulation. These factors along with a high series resistance in the wires will reduce available spark voltage. Trouble of this nature may be difficult to locate since the symptoms are the same as for spark-plug malfunction. An ohmmeter is an excellent instrument for checking electrical continuity of wires and measuring resistance to be sure that it has not become excessive.

Resistor Spark Plugs

With some vehicles, impedance or resistance wires do not completely suppress the RFI. It may be necessary in addition to use resistor-type spark plugs. These plugs have 5000- to 10,000-ohm resistors built into the spark-plug stem. These resistors suppress the radiation that would otherwise occur from the spark plug itself.

Radio Frequency Interference Shield

Introduced on the 1970 GM V-8 models was the two-piece radio frequency interference shield covering the distributor points and secured to the breaker plate by two screws. The function of the shield is to further assist in the reduction of television and radio interference, particularly on GM's windshield-mounted radio antenna installations. When replacing breaker points, be sure the point and condenser lead terminals are properly positioned, back to back, and are firmly secured to the connector. The terminals should then be bent slightly toward the cam to relieve the possibility of the terminals shorting out against the inside of the shield. Either slip-type or screw-type fastener breaker points may be used, providing sufficient clearance exists between the point set connector and the inside surface of the shield. Before tightening the shield mounting screws, be sure the primary ignition or condenser leads are not caught under the edge of the shield. The use of a combination point and condenser set (uniset) eliminates the use of the RFI shield.

Increased Rotor Gap

The increased rotor gap is another means employed to reduce television and radio interference. The rotor gap prior to 1969 was approximately 0.025 inch and the increased gap is about 0.075 inch. This design change makes it increasingly important to use the correct parts when replacing rotors and caps. Be sure to apply the increased kilovolt (kV) rotor gap specification to these installations if you are testing with an oscilloscope to avoid condemning good caps and rotors. The smaller-gap kilovolt specification was approximately 3 kilovolts; the increased gap specification is approximately 8 kilovolts.

Notes

Figure No. 75

CYLINDER NUMBERING SEQUENCES

IN-LINE ENGINES

4-CYLINDER   6-CYLINDER

HORIZONTALLY-OPPOSED

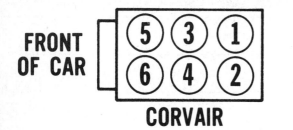

CORVAIR   VOLKSWAGEN AND PORSCHE

V - TYPE

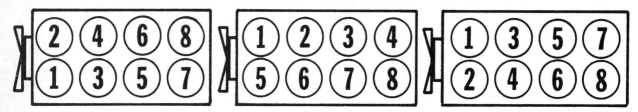

V-8

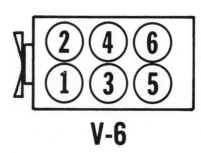

V-6

© Copyright 1986, Tune-Up Manufacturers Institute

CYLINDER-NUMBERING SEQUENCE AND FIRING ORDERS

Every engine has both a cylinder-numbering and a firing-order sequence. The cylinder numbers identify the cylinders according to their location in the engine. The cylinders are usually numbered from front to rear. The firing order of an engine is the listing of the cylinder numbers in the sequence in which they are fired. There are three general arrangements of cylinders in automotive engines: in-line, V-type, and horizontally opposed.

In the in-line design the cylinders are positioned one behind the other. This cylinder arrangement is typical of most four-and six-cylinder engines.

The V-type engine, popular in V-8 engine design, has four cylinders set in each of two banks with the banks set 90 degrees apart. A V-6 engine employs the V-type design with three cylinders in each bank.

The horizontally opposed type of engine design has two engine cylinder banks set 180 degrees apart, giving the engine a flat appearance. The Chevrolet Corvair and the German Volkswagen and Porsche imports use engines of this design.

Regardless of the design of an engine or the number of cylinders it contains, the pistons in the engine move in pairs. That is, there are always two pistons which are attached to crankpins having a common centerline. The engine is designed in this manner to provide the necessary mechanical balance and because all the cylinders in an engine must be fired in two revolutions (720 degrees) of the crankshaft.

The location of cylinder 1 is important to the tune-up specialist because it is a reference point for ignition timing and for valve timing. Cylinder 1 is usually the cylinder nearest the radiator. In the case of V-type engines however, cylinder 1 may be in either bank. The engine diagram in Figure No. 75 illustrates the cylinder 1 position on popular production engines. Refer to your tune-up specifications for both the cylinder-numbering sequence and the firing order of the engine you are tuning.

Figure No. 76

SPARK PLUG HEAT RANGE

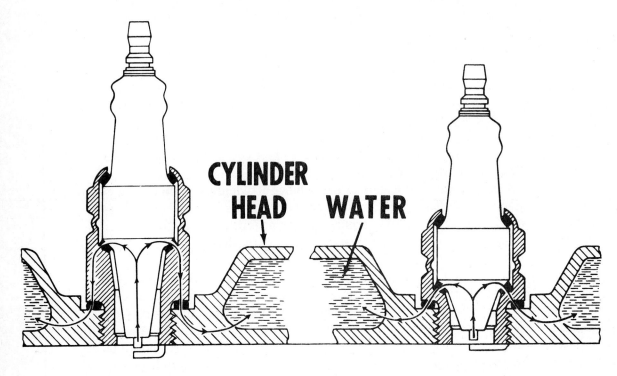

HOT PLUG — COLD PLUG

SPARK PLUG REACH

PROPER REACH

REACH TOO SHORT

REACH TOO LONG

© Copyright 1986, Tune-Up Manufacturers Institute

SPARK-PLUG HEAT RANGE

To provide proper engine performance, spark plugs must operate within a certain temperature range. If the spark plugs consistently operate at too cold a temperature, less than 700°F, soot and carbon will deposit on the insulator tips, which cause fouling and missing. If the plugs run too hot, more than 1700°F, the insulator will be damaged, and electrodes will burn away rapidly. In extreme conditions, hot plugs cause premature burning (preignition) of the air-fuel mixture.

The ability of a spark plug to transfer heat from the insulated center electrode tip is controlled by the design of the spark plug. The only path for heat to escape is through the insulator tip, spark-plug shell and gasket, through the cylinder head, to the cooling liquid in the water jacket. By varying the length and shape of the insulator and shell, the manufacturer is able to produce spark plugs with different heat-range characteristics and thereby control their operating temperatures.

A visual inspection of the spark plugs after they are removed from the engine will reveal the existence of a variety of engine ailments in addition to indicating, with reasonable accuracy, the correctness of the heat range of the spark plugs used. The nature of the deposits collected on the plug insulator and the condition and the gap of the electrodes also provide valuable clues. It is important, however, not to idle a cold engine for any length of time before removing the spark plugs for examination. The plugs taken from a cold engine that has been idling with partial choke will very likely be soot fouled by the rich fuel mixture. This deposit will give a false impression of the true condition of the plug.

When replacing spark plugs during your tune-up, it is usually advisable to replace the plugs with the same heat range. If, however, the plugs in an engine constantly exhibit electrode wear and blistered insulators, a colder range of plugs should be used. If, because of constant low-speed city driving, the plugs are constantly fouled, a hotter range of plugs should be installed. When a change is made, change only one heat range number at a time.

Remember, a hotter spark plug has a higher temperature, not a hotter spark.

Spark-Plug Reach

Spark-plug reach is the length of the threaded portion of the plug. If the reach is too short, the plug electrodes will be in a pocket and may misfire under certain conditions. The exposed threads in the cylinder head will "carbon up," making cleaning of the threads necessary before plugs of the proper reach can be correctly installed and torqued. If the reach is too long, the plug threads will be exposed and may overheat, resulting in preignition. The exposed threads will also carbonize, making plug removal difficult. The danger of piston head damage also exists if the plug reach is too long.

Figure No. 77

SPARK PLUG FEATURES

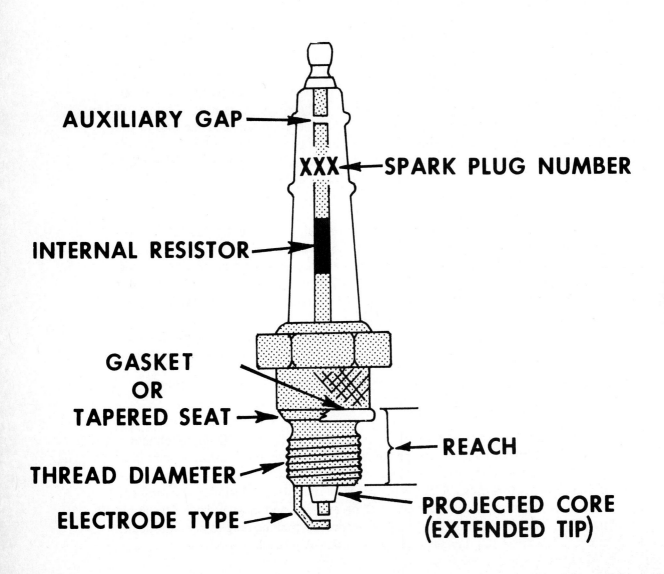

SPARK-PLUG FEATURES

Besides the important heat-range and reach requirements of the spark plug, there are several other features that are highly instrumental in proper spark-plug performance. Since every feature is important in proper plug selection, the tune-up specialist should understand what the plug markings represent.

To the basic requirements of heat range and reach are added special design features, such as: tapered seat; internal series booster gap; internal resistor; two ground electrodes; projected firing tip; heavy-duty electrodes and other features.

To assist the tune-up specialist in the proper selection of spark plugs for the engine he is tuning, the plug manufacturers mark their plugs with numbers and letters to identify every specification and every feature. The following are only representative examples of these markings.

Champion

Champion spark-plug markings have the following meanings:

1. The prefix (first) letter indicates the thread diameter and plug reach. Examples of popular application are J is 14-millimeter (mm) thread, 3/8-inch reach; L is 14-mm thread, 1/2-inch reach; N is 14-mm thread, 3/4-inch reach; and F is 18-mm thread with tapered seat.
2. If the plug has any special design features, a prefix letter indicating the feature will be placed before the first prefix letter. *Example:* V is a tapered seat and R is a built-in resistor.
3. The number following the prefix letter(s) is the heat range. If the range of numbers is, for example, 3 to 18, then plug 3 is the coldest and plug 18 is the hottest. In between these two numbers is a gradual increase in heat range with every increase in number.
4. The suffix (last) letter indicates the spark-gap design. *Examples:* B is a two-ground electrode; Y is a projected core nose; C is a copper-core center electrode.
5. The suffix (last) number indicates electrode gap width. *Example:* −4 or −6 could be 0.040- to 0.060-inch gap.
6. RJ12YC. This plug has a resistor (R), has a 14-mm thread with a 3/8-inch reach (J), is No. 12 (12) in the heat range, and has a projected core nose (Y) with a copper-core center electrode (C).
7. RV12YC. This plug has a resistor (R), has a tapered seat (V) and a 14-mm thread with a 1/2-inch reach, is No. 12 (12) in the heat range, and has a projected core nose (Y), with a (C) for copper-core center electrode.
8. RJ-11. This plug has a built-in resistor (R), has a 14-mm thread with a 3/8-inch reach (J), and is No. 11 (11) in the heat range.
9. RV15YC4. This plug has a resistor (R), has a tapered seat (V) and a 14-mm thread with a 1/2-inch reach, is No. 15 (15) in the heat range, with a copper-core center electrode (C), and has an electrode gap width of 0.040 inch (−4).

© Copyright 1986, Tune-Up Manufacturers Institute

AC

The spark plugs manufactured by the AC Division of General Motors Corporation are identified in a somewhat similar manner:

1. The prefix (first) letter indicates the spark plug features. Example: B is a series gap; C is commercial (heavy-duty) electrodes; R is a built-in resistor.
2. Following the prefix letter are two numbers. The first of the two numbers indicates the thread size. All numbers starting with 2 are 1/2 inch; all numbers starting with 4 are 14 mm; all numbers starting with 7 are 7/8 inch; among others. The second number indicates the heat range. The higher the number, the hotter the plug. A 46 plug (6) is hotter than a 44 plug (4), although both plugs have a 14-mm thread by virture of the first No. 4.
3. Suffix letters following the numbers indicate special design features. Examples: XL is extralong reach; S is extended tip; T is tapered seat; and TS is tapered seat with extended tip.

Following are a few AC spark-plug numbers using the above data:

1. 45S. This plug has a 14-mm (4) thread; is No. 5 in the heat range (5); and has extended tip (S).
2. 44TS. This plug has a 14-mm (4) thread; is No. 4 in the heat range (4); and has a tapered seat with an extended tip (TS).
3. 45XL. This plug has a 14-mm (4) thread; is No. 5 in the heat range (5); and has an extralong 3/4-inch reach (XL).

Autolite

On Autolite spark plugs the first letter is the thread size. Examples: A is a 14 mm; B is an 18 mm.

1. The second letter is the reach. Examples: None is 3/8 inch; L is 7/16 inch; E is 1/2 inch; and G is 3/4 inch.
2. If there are other second or third letters, they represent other special design features. Examples: R is a built-in resistor; F is a tapered seat.
3. Following the prefix letter(s) are numbers that represent the heat range. The numbers run from 2, which is the coldest plug, to 11, which is the hottest. Autolite classifies their heat-range numbers into three catagories. Numbers 2 through 5 are cold plugs. Number 6 through 8 are medium plugs. Numbers 9 through 11 are the hot plugs.
4. Letters after the heat-range numbers are special features. Example: S is for shield; X is for outboard marine; and M is for moisture-proof pack.

Typical examples of Autolite spark plugs using the above symbols are as follows:

1. A5. This plug has a 14-mm thread (A); has a 3/8-inch reach (because of no number); and is No. 5 in the heat range (cold).
2. AE6. This plug has 14-mm thread (A); has a 1/2-inch reach (E); and is No. 6 in the heat range (medium).
3. AER6. This plug has a 14-mm thread (A); has a 1/2-inch reach (E); has an internal resistor (R); and is No. 6 in the heat range (medium).

Prestolite

Prestolite spark-plug markings have the following meanings:

1. Plug diameter: first two numerals indicate thread diameter. Examples: 10 is 10 mm; 12 is 12 mm; 14 is 14 mm; 18 is 18 mm.
2. Plug reach: the following letters indicate reach. L is 7/16 inch; E is 1/2 inch; G is 3/4 inch.
3. Spark-plug type: the following letters indicate design types. R is resistor; F is tapered seat; O is surface gap; X is small engine; N is miniplug; A is wide gap.
4. Heat range: third and, if necessary, fourth numerals indicate heat range. For example, 9 through 11 are of hot range; 6 through 8 are of medium range; 1 through 5 are of cold range.
5. Thermal tip: spark plugs with thermal tips are designated by digit 2 as the final numeral.

Example: 14GR52A. 14 indicates 14 mm; G indicates 3/4-inch reach; R indicates resistor; 5 indicates heat range; 2 indicates thermal tip; A indicates wide gap.

It can be readily seen from the few examples cited that there is a lot more to the proper selection of a set of spark plugs than merely "picking a number." At best, the wrong plugs will result in poor engine performance and short plug life; at worst, they can cause severe engine damage for which the mechanic could be held responsible.

Figure No. 78

CAPACITOR DISCHARGE IGNITION SYSTEM

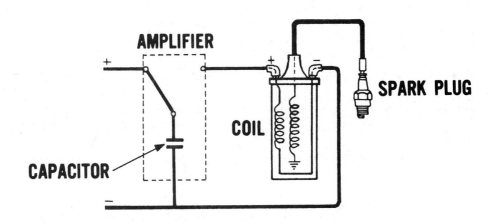

CAPACITOR BEING CHARGED

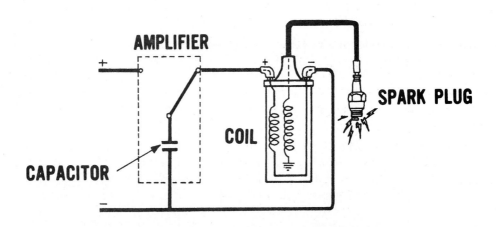

CAPACITOR DISCHARGING INTO COIL PRIMARY WINDINGS

CAPACITIVE DISCHARGE IGNITION SYSTEM

Two types of capacitive discharge (CD) ignition systems exist. One type uses breaker points as a switching function, the other a magnetic impulse device which replaces breaker points. The CD systems using breaker points are designed to decrease the amount of current flowing through the points from about 4 to 6 amps to about 2 amps so that they will wear less, last longer, and eliminate the need for a condensor. These units use points only for a switching function and not to carry primary current flow. Dwell angle and saturation time are not dependent on the points, but are controlled by the circuitry built into the electronic control unit.

Capacitive discharge ignition systems operate using a high-voltage capacitor (condenser) connected across the coil primary windings to the control unit. During the time the spark plugs are not firing (normal dwell time), the capacitor is charged with approximately 300 to 400 volts. When either the breaker points open or the magnetic device sends an impulse signal, the capacitor discharges or dumps its high-voltage charge into the coil primary windings. The step-up transformer action of the ignition coil increases this primary voltage to as much as 35,000 volts on the demand of the ignition system. By maintaining maximum primary circuit output, complete coil saturation and maximum secondary circuit output voltage are accomplished.

The CD ignition system is said to possess several advantages over other ignition systems. In systems other than CD, the primary circuit voltage is reduced because of varying battery voltage, due to charge and discharge cycles or lack of sufficient charging system output, when full electrical system demands are required. This, of course, results in a proportionate loss in secondary voltage output.

In the CD system, the amplifier will load the capacitor with a full charge of 300 to 400 volts if the battery only has a sufficient charge to turn the engine over. Another desirable factor is that the high-voltage spark is delivered in less than normal time. High-voltage leakage that occurs over distributor-cap surfaces, moisture-covered spark-plug cables, and fouled or gasoline-dampened spark plugs is thereby minimized. The combination of consistent high-voltage impulses and fast spark delivery is said to make the cold-weather starting of even a flooded engine much easier.

In addition to good engine starting capability, the constant primary circuit input to the coil assures maximum voltage output through the vehicle's entire speed range. There will be no "running out of ignition" at high speed. On units where the distributor is breakerless, there will be no voltage fall-off due to high-speed breaker point bounce. Spark plugs with wider than normal gaps (resulting from high mileage) also will be consistently fired by the CD system.

The mechanical switch, shown within the broken line in Figure No. 78, is for explanatory purposes only. The actual charging of the capacitor and the switching from system charge to discharge is accomplished electronically by transistors, diodes, resistors, capacitors, a transformer, and other units in the amplifier.

Notes

7

ELECTRONIC IGNITION SYSTEMS

CHRYSLER ELECTRONIC IGNITION SYSTEM (EXCEPT LEAN BURN)

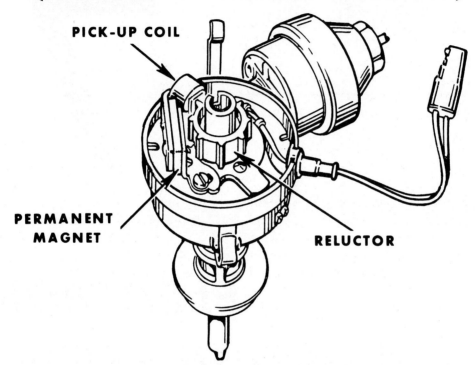

DISTRIBUTOR COMPONENTS

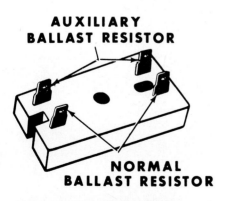

BALLAST RESISTOR

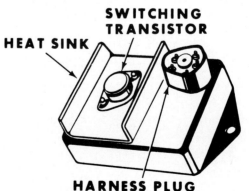

ELECTRONIC CONTROL UNIT

CHRYSLER ELECTRONIC IGNITION SYSTEM (EXCEPT LEAN BURN)

Chrysler's Electronic Ignition System is designed for use on all their 1972 and later engines. The system is breakerless, thus eliminating the need for both breaker points and condenser.

Two short primary leads running to the distributor from a quick disconnect provide an easy method of identifying the ignition system. A dual-type ballast resistor (or single on later models) is mounted on the firewall, and an electronic control unit containing a switching transistor is positioned on either the firewall or fender shield. A standard ignition coil along with the necessary wiring harness complete the components of the system.

The distributor plate contains a permanent magnet and a pick-up coil instead of the conventional breaker points and condenser. The distributor cam is replaced with a four-, six-, or eight-tooth reluctor.

The basic electrical law on which the electronic circuit functions is that, anytime a magnetic field is broken, a proportionate voltage is induced. Each time one of the teeth of the distributor shaft-driven reluctor passes the permanent magnet, the magnetic field is interrupted and a voltage signal or "timing pulse" is induced in the pick-up coil and sent to the control unit. The switching transistor then interrupts the primary circuit, thereby generating high voltage in the coil secondary windings to fire the spark plugs at the precise instant required.

The switching transistor interrupts the primary circuit for a specific length of time dictated by the timing pulse created by the reluctor and pick-up coil. This primary circuit interruption constitutes the dwell angle. The dwell angle can be read with a dwell meter, but the angle is not adjustable. No provision for dwell-angle adjustment has been provided since the design of the reluctor precludes the possibility of any dwell-angle change. Variations in ignition timing, subsequent to changes in dwell-angle setting, have also been eliminated. Chrysler states that periodic checks of dwell-angle settings are not necessary on the electronic ignition system.

Preventative maintenance of the electronic ignition system is reduced to occasional inspection and testing of the spark-plug cables, distributor cap and rotor and replacement of the spark plugs, as required. In the event the magnet or pick-up coil become defective, their replacement can be performed as a complete unit. Other system units can be similarly tested and replaced with ease.

CHRYSLER ELECTRONIC LEAN BURN SYSTEM

Figure No. 80

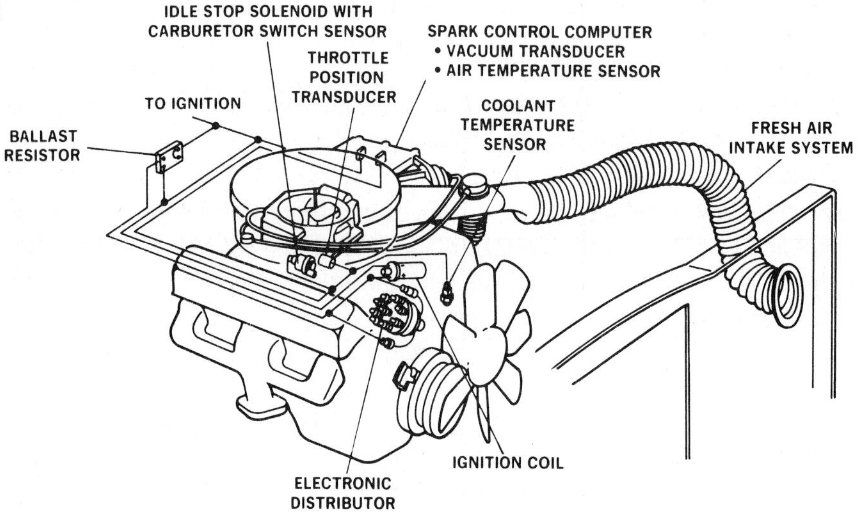

Figure No. 81

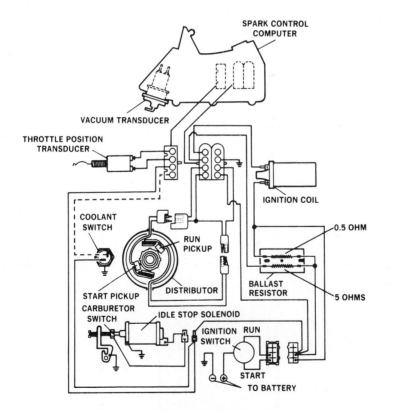

CHRYSLER ELECTRONIC LEAN BURN SYSTEM

The electronic lean burn system consists of a spark control computer, various engine sensors, and a specially calibrated carburetor. The function of the system is to provide a way for the engine to burn a lean air-fuel mixture.

Spark Control Computer

The spark control computer is the heart of the entire system. It provides the capability of igniting a lean fuel mixture in relationship to different modes of engine operation by delivering an infinite amount of variable advance curves. The computer consists of two electronic printed-circuit boards. They are the *program schedule module* and the *ignition control module.* They work together as follows: The program module simultaneously receives signals from all sensors and within milliseconds computes them to determine how the engine is operating and then directs the ignition module to advance or retard the firing of the spark plugs accordingly. It must be understood that this advancing and retarding of the ignition timing is not based on a constant curve. As mentioned before, the curves can be infinite and variable.

Sensors

There are seven sensors on the engine which supply the spark control computer with the necessary information needed to fire the spark plugs at the right time. They, and their functions, follow:

Start

Located in the distributor, the start pick-up supplies the signal to the ignition control module of the computer, which will cause the spark plugs to fire at a fixed amount of advance during cranking.

Run Pick-Up

Located in the distributor, the run pick-up supplies the basic timing signal to the computer. This signal will tell the computer to create the maximum amount of timing advance available for any engine rpm. The computer can also determine engine speed and when each piston is coming up on its compression stroke from this signal.

Coolant-Temperature Sensor

Located on the water-pump housing, the coolant-temperature sensor signals the computer when engine coolant temperature is below 150°F.

Air Temperature Sensor

Located inside the computer, the air temperature sensor supplies a signal to the computer, which tells the temperature of the air coming in the cleaner from the fresh-air system. This signal also affects the amount of additional spark advance given by the computer that is going to be created by the throttle position transducer signal.

Throttle Position Transducer

Located on the carburetor, the throttle position transducer signals the computer the position and the rate of change of the throttle plates. It works by changing mechanical motion into an electrical signal which the spark control computer uses to control additional spark advance. Additional spark advance will be given by the computer when the throttle plates start to open, and in every position to full throttle. Even more advance is given for about 1 second if the throttle is opened quickly. However, as mentioned before, the air temperature-sensor signal is going to control the maxiumum allowable advance. If the air temperature in the air cleaner is hot, less advance will be allowed. On the other hand, if it is cold, more advance will be allowed.

Carburetor Switch Sensor

Located on the carburetor, the carburetor switch sensor tells the computer if the engine is at idle or off idle.

Vacuum Transducer

Located on the computer, the vacuum transducer tells the computer what intake manifold vacuum is. The higher the vacuum, the more additional advance will be given. The lower, the less amount of advance. To obtain the maximum amount of advance for any inch of vacuum, the carburetor switch sensor must remain open for a specified amount of time. During that time the advance will not happen quickly but will build up at a slow rate. If the carburetor switch closes before the predetermined time period, the buildup of advance at that time will be canceled in the ignition system; however, the computer will put it into memory and slowly return it to zero. If the switch is reopened before the advance is returned to zero, the buildup of advance starts at the point where the computer still has it in memory. If the switch is reopened after the advance is returned to zero, the buildup of advance must start all over again.

System Operation

When the ignition key is turned to start position, the start pick-up in the 1976-77 system will send its signal to the ignition module to the spark control computer, and additional spark advance will be given during cranking to assist in engine starting. Starting in 1978, the system no longer has a start pick-up. Immediately after the engine starts, the run pick-up takes over and sends its signal to the computer. Also, the computer will create additional advance and will maintain it for approximately 1 minute. However, during that time period the additional advance will slowly be eliminated. With the engine running, and if the engine coolant temperature is below 150°F, the coolant-temperature sensor will signal the computer of this to prevent any additional spark advance to occur from the vacuum transducer signal. After the engine reaches operating temperature, normal system operation will begin.

The run pick-up sends its basic timing signal to the computer, which creates the maximum amount of timing advance available for any engine rpm. At the same time, input signals from the air temperature sensor, throttle position transducer, carburetor switch sensor, and vacuum transducer are received by the computer and are calculated to determine how much of the total advance is required in the firing of the spark plugs to meet the engine's operating conditions.

Finally, if for some reason there is a failure of the computer or the run pick-up coil, the system will go into what is called the *limp in mode*. This will enable the driver to continue to drive the vehicle until it can be repaired. However, while in this mode very poor performance and fuel economy will be given by the system. If there is a failure of the start pick-up or ignition control module of the computer, the engine will not start on run.

TROUBLESHOOTING THE CHRYSLER ELECTRONIC IGNITION SYSTEM

Figure No. 82

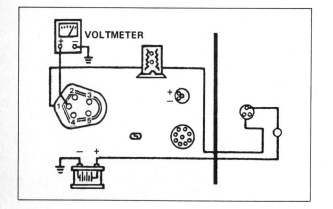

Figure No. 83

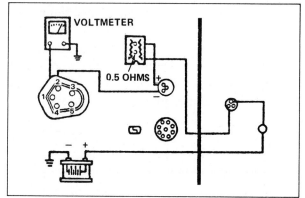

Figure No. 84

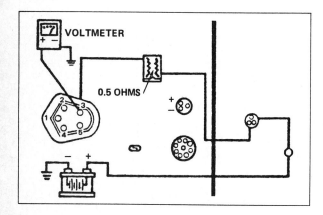

Figure No. 85

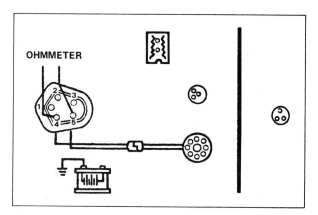

Figure No. 86

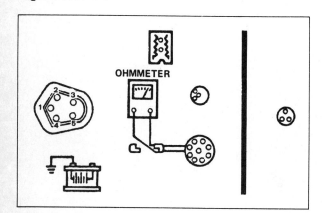

Figure No. 87

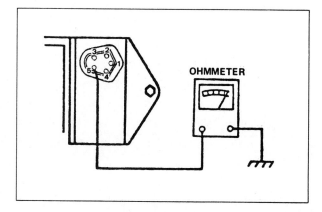

© Copyright 1986, Tune-Up Manufacturers Institute

TROUBLESHOOTING THE CHRYSLER ELECTRONIC IGNITION SYSTEM

1. Check battery connections and voltage at battery.
2. Remove distributor cap; check air gap between reluctor and pick-up coil tooth. (Should be 0.008.) Make necessary adjustment with a nonmagnetic feeler gauge.
3. Make a visual inspection of all secondary cables; look for cracks on spark-plug boots or on distributor boots. Also check the primary wires at the coil.
4. Turn all accessories off; *makes sure ignition switch is off.*
5. Remove multiwiring connector from control unit.
6. Connect negative lead of voltmeter to a good ground.
7. Turn ignition switch on—check available voltage at cavity 1 on wire harness (see Figure No. 82). Voltage should be within 1 volt of battery voltage. If there is more than 1-volt loss, check circuit outlines in Figure No. 82 for loose or broken connections.
8. Check available voltage at cavity 2; available voltage should be within 1 volt of battery voltage; if not, check circuit outlines in Figure No. 83.
9. Check available voltage at cavity 3; available voltage should be within 1 volt of battery voltage; if not, check circuit outlines in Figure No. 84.
10. Turn ignition switch off.
11. Now take an ohmmeter and check cavities 4 and 5. The ohmmeter reading should be between 150 and 900 ohms (see Figure No. 85). If reading is higher or lower, disconnect dual lead from distributor. Check resistance of pick-up coil (see Figure No. 86). If reading is not between 150 and 900 ohms, replace pick-up coil. If reading is within specifications, check the wiring harness from dual lead connector back to control unit. Connect ohmmeter to a good ground and the other lead to either dual lead connectors pin of the distributor harness. Ohmmeter should show an open circuit. If the ohmmeter shows continuity, the pick-up coil in the distributor must be replaced.
12. Connect one ohmmeter lead to a good ground and the other lead to the *control unit* pin 5. The ohmmeter should show continuity between ground and the connector pin (see Figure No. 87). If continuity does not exist, tighten bolts holding control unit. Now recheck. If continuity still does not exist, control unit must be replaced.
13. Reconnect wiring harness at control unit and distributor. Remove coil wire from center of distributor. Hold approximately 3/16 inch from a good ground.

Important: Never allow the coil to operate open circuited, that is, with the coil wire completely out of the coil tower. The high voltage may possibly cause damage to the solid-state control unit, especially to the transistor.

Crank engine. If arcing does not occur, the control unit or the coil will have to be replaced.

14. Before replacing control unit, perform a coil check. It is very unlikely that both the control unit and the coil will both fail at the same time. Check ignition coil with an ohmmeter. The primary reading at 70 to 80°F should be 1.41 to

1.79 ohms. At the same temperature span the secondary reading should be 9200 to 11,700 ohms. Check for carbon tracking on the coil tower. If detected, replace.

15. Use a good quality ohmmeter to test the ballast resistor. The auxiliary resistor is rated at 4.75 to 5.75 ohms of resistance. The normal resistor which is wire wound should read 0.5 to 0.6 ohms. *Note:* If these specifications are not met by either resistor or is open circuit, the unit is defective and must be replaced.

Notes

Figure No. 88

DELCO HIGH—ENERGY IGNITION (HEI) SYSTEM

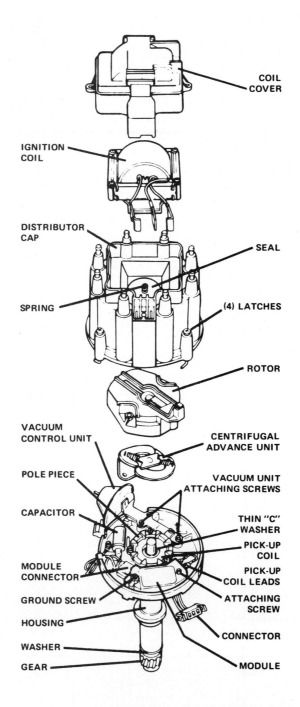

DELCO HIGH-ENERGY IGNITION (HEI) SYSTEM

Delco-Remy introduced their high-energy ignition (HEI) system in 1974. It is similar in operation to its predecessor, the unitized ignition system, but has several physical differences. Although larger, the distributor cap is similar in appearance to a conventional one. The diameter of the cap has been enlarged to prevent arcing between the terminals. The secondary wires are no longer molded into the cap; they are individual 8-mm wires which attach directly to the cap terminals. The ignition coil for the V-8 and V-6 models is in the distributor cap and is in direct contact with the rotor. Up until 1979 the four-and six-cylinder in-line engines had an external coil. Now all coils are in the cap. In addition, a solid-state electronic module together with a magnetic pick-up assembly takes the place of the conventional point and condenser. Other parts include a rotor, vacuum unit, and capacitor, which is used only for radio noise suppression. HEI components are not interchangeable with unitized ignition components.

It is not necessary to remove the V-6 or V-8 cylinder coil cover in order to remove the distributor cap. The cap has four fasteners. Removal is accomplished by depressing and turning these fasteners. The rotor is of conventional design and removed by loosening the screws on both sides.

The magnetic pick-up assembly consists of a rotating timer core attached to the distributor shaft, a pole piece with internal teeth, a permanent magnet, and a pick-up coil. When the teeth of the timer core rotating inside the pole piece line up with the teeth of the pole piece, an induced voltage in the pick-up coil signals the electronic module to trigger the coil primary circuit. The primary current decreases, and a high voltage is induced in the ignition-coil secondary winding, which is directed through the rotor and secondary wires to fire the spark plugs.

The magnetic pick-up assembly is mounted over the main bearing on the distributor housing and made to rotate by the vacuum-control unit, thus providing vacuum advance. The timer core is made to rotate about the shaft by conventional advance weights, thus providing centrifugal advance.

The module automatically controls the dwell period, stretching it with increasing engine speed. The HEI system also features a longer spark duration, made possible by the higher amount of energy available to the secondary and utilizing a wider spark-plug gap.

TROUBLESHOOTING THE DELCO HIGH-ENERGY IGNITION (HEI) SYSTEM

Figure No. 89

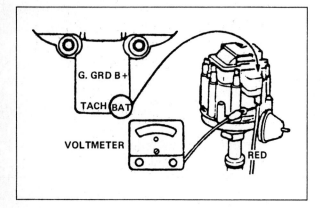

Figure No. 90

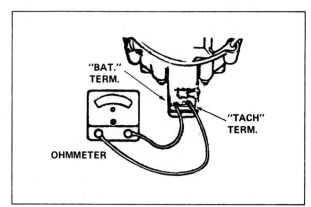

Figure No. 91

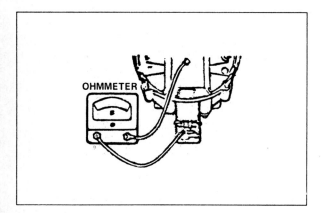

Figure No. 92

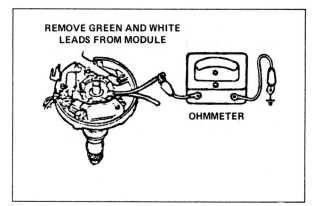

Figure No. 93

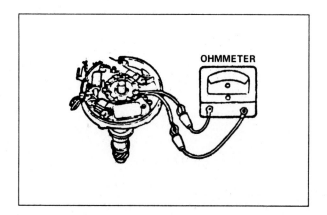

TROUBLESHOOTING THE DELCO HIGH-ENERGY IGNITION (HEI) SYSTEM

Important: When performing cylinder compressor test disconnect the ignition switch connector from the HEI system on the side of the unit.

Engine Will Not Start

1. Check battery connections and voltage of battery. Check distributor connector.
2. Connect a voltmeter (Figure No. 89) between BAT terminal lead on the distributor and a good engine ground. Turn ignition switch to the run position.
   a. If a zero reading is obtained, locate and repair the open circuit between the BAT terminal and the battery.
   b. If the voltage reading is equal to battery voltage, hold one spark plug lead about ¼ inch from a good engine ground. Crank engine with the starter. If no spark occurs, then proceed with the component checks.
   c. If spark occurs, the trouble is not in the ignition system.

Component Checks

1. Remove the distributor cap and coil assembly; inspect for cracks or rotor arcing.
2. On three-wire style coils (1974 and mid-year 1975 models) connect ohmmeter leads to the two male connectors in the distributor cap. On four-wire-style coils connect ohmmeter to center ground terminal on cap and to rotor button inside the cap (use the high scale).
   a. A reading of zero or near zero should be obtained. If not, replace the coil (Figure No. 90).
3. Connect the ohmmeter from the tach terminal on the cap to the rotor button inside the cap (use the high scale).
   a. If a reading is more than 30,000 or less than 6,000 ohms, replace the coil.
4. Connect a test vacuum source to the vacuum unit. If the vacuum unit is inoperative, replace it.
5. Detach the pick-up coil leads to the module and connect the ohmmeter to one of the pick-up coil leads; connect other lead to a good engine ground (use x 1000 scale). (Figure No. 91).
   a. If meter reading is less than 500 ohms or more than 1500 ohms, replace the pick-up coil. Apply vacuum—observe the reading throughout the vacuum range.
   b. If it reads other than infinite at anytime, replace the pick-up coil (Figure No. 92).
6. Connect the ohmmeter leads to pick-up coil leads.
   a. If meter reading is less than 500 ohms or more than 1500 ohms, replace the pick-up coil.
7. If no defects have been found, replace the module.

© Copyright 1986, Tune-Up Manufacturers Institute

Engine Runs Rough

8. Check the fuel system and make sure the right amount of fuel is being delivered to the carburetor. Check all hoses and fittings for leakage.
9. Visually inspect the system and make sure arcing is not occurring.
10. Check timing. Set to factory specifications.
11. Check centrifugal advance of the distributor.
12. If no problem exists with spark plugs or spark-plug wires, follow the procedure to component checks.
13. To replace pick-up coil.
    a. Remove unit from engine, remove roll pin from shaft, and remove gear.
    b. Remove rotor and shaft assembly from housing.
    c. Remove C washer.
    d. Remove pickup coil assembly.
    e. Reverse procedure to reassemble.

Notes

FORD SOLID-STATE IGNITION SYSTEM

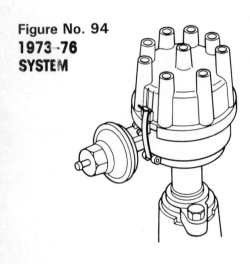

Figure No. 94
1973-76 SYSTEM

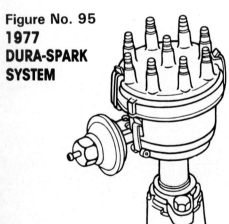

Figure No. 95
1977 DURA-SPARK SYSTEM

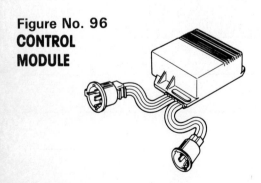

Figure No. 96
CONTROL MODULE

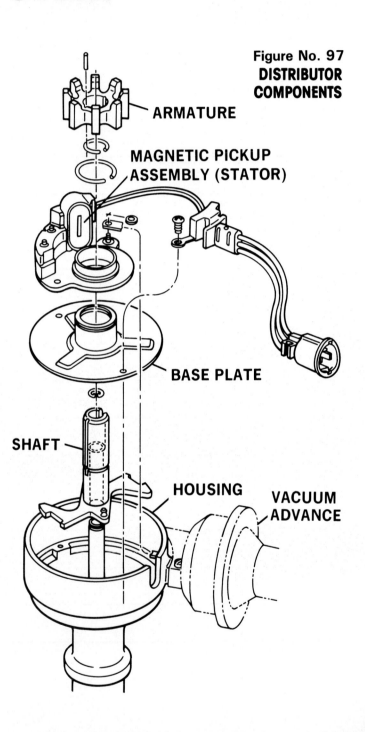

Figure No. 97
DISTRIBUTOR COMPONENTS

- ARMATURE
- MAGNETIC PICKUP ASSEMBLY (STATOR)
- BASE PLATE
- SHAFT
- HOUSING
- VACUUM ADVANCE

FORD SOLID-STATE IGNITION SYSTEM (PRE-1977, DURA-SPARK I, II)

Ford's electronic ignition system first became available on some late 1973 models and was standard equipment on all Ford passenger cars and light trucks in 1975.

Generally, Ford's electronic ignition works like the conventional breaker-point system. Ignition spark is generated in a coil similar to that used on earlier models. The system is breakerless by using a magnetic pick-up assembly—stator and rotating armature—to control primary current flow through an electronic control module. When one of the edges of a tooth on the armature is aligned with the stator, an impulse is sent to the control module, which in turn opens the coil primary circuit.

The breakerless system eliminates the breaker-point assembly, condenser, and its inherent problems—wear, resistance, changes in dwell angle, and timing. Electronic ignition insures higher spark output and precise ignition timing over a greatly extended period of time.

The distributor cap, rotor, spark-plug wires, and spark plugs will require periodic inspection, adjustment, or replacement. The electronic parts—including armature, stator, and control module—can be easily replaced if a problem develops.

Troubleshooting procedures for the Ford electronic ignition, found in Chapter 7, are different depending on the model year of the vehicle. The basic difference from 1974 through 1976 models involves the control-module harness socket configurations.

On 1977 models, Ford introduced the Dura-Spark system with enlarged cap, rotor, and 8-mm spark-plug wires. Eight-cylinder, 1977 Ford products, manufactured for use in California, have a Dura-Spark I ignition system. Similar Ford products manufactured for other states have a Dura-Spark II ignition system. The Dura-Spark I system incorporates a new control module. Dura-Spark II systems have the same control module used on 1976 models.

Control modules for all year models are identified by the color of the grommet securing the wiring harness to the module housing.

Caution: When making a spark intensity test, *never* pull off
- No. 1 or No. 8 wire on V-8 engines
- No. 3 or No. 5 wire on I-6 engines
- No. 1 or No. 4 wire on V-6 engines
- No. 1 or No. 3 wire on 4-cylinder engines

Doing so may cause internal arcing within the distributor and burn out the electronic module.

TROUBLESHOOTING THE FORD SOLID-STATE IGNITION SYSTEM (PRE-1977)

Figure No. 98

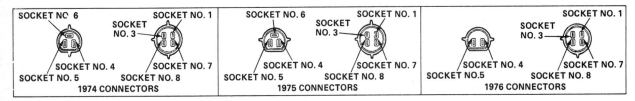

Circuit Checks

1. The following voltage and resistance test checks inputs to the module.
2. The module *must* be disconnected from the harness for all tests except the ignition coil Bat. terminal to ground voltage test. Make all these tests at the *harness* side of the *module* connectors, except the one made at the coil tower.

Ignition Switch Position	Test Voltage Between			Should Be	If Reading Is Incorrect
	1974	1975	1976		
Key on	Ignition coil bat. terminal and engine ground (module connected)			4.9 to 7.9 volts	Less than spec.—check primary wiring. More than spec.—check resistance wire.
	Socket 3 and engine grd.	Socket 4 and engine grd.	Socket 4 and engine grd.	Battery voltage ± 0.1 volts	Check supply wire and connectors through ignition switch.
	Socket 5 and engine grd.	Socket 1 and engine grd.	Socket 1 and engine grd.	Battery voltage ± 0.1 volts	Check wire to ignition coil and/or ignition coil.
Cranking	Socket 1 and engine grd.	Socket 5 and engine grd.	Socket 5 and engine grd.	8 to 12 volts	Check supply wire and connectors through ignition switch.
	Socket 5 and engine grd.	Socket 6 and engine grd.	Socket 1 and engine grd.	8 to 12 volts	Check solenoid by-pass circuit.
	Socket 7 and Socket 8	Socket 7 and Socket 3	Socket 7 and Socket 3	½ volt minimum ac or any dc volt variation	Perform hardware test (see below).
	Test Resistance Between			Should Be	If Reading Is Incorrect
Key off	Socket 7 and Socket 8	Socket 7 and Socket 3	Socket 7 and Socket 3	400 to 800 ohms	
	Socket 6 and engine grd.	Socket 8 and engine grd.	Socket 8 and engine grd.	0 Ohm	If any test fails, first check for defective harness to distributor.
	Socket 7 and engine grd.	Socket 7 and engine grd.	Socket 7 and engine grd.	70,000 ohms or more	If harness checks good, distributor stator assembly must be replaced.
	Socket 8 and engine grd.	Socket 3 and engine grd.	Socket 3 and engine grd.	70,000 ohms or more	
	Socket 3 and coil tower	Socket 4 and coil tower	Socket 4 and coil tower	7000 to 13,000 ohms	Check coil on coil tester.
	Ignition coil primary terminals			1.0 to 2.0 ohms	Check coil on coil tester.
	Socket 5 and engine grd.	Socket 1 and engine grd.	Socket 1 and engine grd.	More than 4.0 ohms	Check for short to ground at DEC terminal of coil or in primary coil winding.

Distributor Hardware Test

1. Disconnect three-wire weatherproof connector at distributor pigtail.
2. Connect a dc voltmeter on 2.5-volt scale to two parallel blades. With the engine cranking, the meter needle should oscillate.
3. Remove distributor cap and check for visual damage or misassembly.
   a. Sintered iron armature (four-, six-, or eight-toothed wheel) must be tight on sleeve, and the roll pin aligning armature to sleeve must be in position.
   b. Sintered iron stator must not be broken.
   c. Armature must rotate when engine is cranked.
4. If the hardware is OK, but the meter doesn't oscillate, replace the magnetic pick-up assembly.

If above checks are OK, substitute a known good module. If this corrects malfunction, reconnect original module to verify that connector is not causing problem.

TROUBLESHOOTING THE FORD SOLID-STATE IGNITION SYSTEM (PRE-1977)

Circuit Checks

1. The following voltage and resistance test checks inputs to the module.
2. The module *must* be disconnected from the harness for all tests except the ignition coil BAT terminal to ground voltage test. Make all these tests at the *harness* side of the *module* connectors, except the one made at the coil tower.

Distributor Hardware Test

1. Disconnect three-wire weatherproof connector at distributor pigtail.
2. Connect a dc voltmeter on 2.5-volt scale to two parallel blades. With the engine cranking, the meter needle should oscillate.
3. Remove distributor cap and check for visual damage or misassembly.
   a. Sintered iron armature (four-, six-, or eight-toothed wheel) must be tight on on sleeve, and the roll pin aligning armature to sleeve must be in position.
   b. Sintered iron stator must not be broken.
   c. Armature must rotate when engine is cranked.
4. If the hardware is OK, but the meter doesn't oscillate, replace the magnetic pick-up assembly.

If above checks are OK, substitute a known good module. If this corrects malfunction, reconnect original module to verify that connector is not causing a problem.

TROUBLESHOOTING THE FORD SOLID-STATE 1977 IGNITION SYSTEM (DURA-SPARK SYSTEMS I, II)

Figure No. 99   Figure No. 100

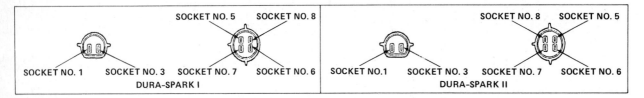

Circuit Checks

1. The following voltage and resistance test checks inputs to the module.
2. The module *must* be disconnected from the harness for all tests except the ignition coil Bat terminal to ground voltage test on the Dura Spark II system. Make all these tests at the harness side of the *module* connectors except the one made at the coil tower.

Ignition Switch Position	Test Voltage Between		Should Be	If Reading Is Incorrect
	Dura-Spark I	Dura-Spark II		
Key on	Test Bat. terminal on coil to eng. grd. Use jumper wire to ground other coil term.		11.0 to 14.0 volts (Dura I) 4.9 to 7.9 volts (Dura II)	Less than Spec.—Check primary wiring. More than Spec.—Check resistance wire.
	Socket 3 and eng. grd.		Battery voltage ± 0.1 volts	Check supply wire and connectors through ignition switch
	Socket 5 and eng. grd.		Battery voltage ± 0.1 volts	Check wire to ignition coil and/or ignition coil.
Cranking	Socket 1 and eng. grd.		8 to 12 volts	Check supply wire and connectors through ignition switch.
	Jumpers 5 and 6—Read bat. coil terminal and eng. grd.		8 to 12 volts	Check solenoid by-pass circuit.
	Sockets 7 and 8		½ volt minimum ac or any dc volt variation	Perform hardware test (see below).
	Test Resistance Between		Should Be	If Reading Is Incorrect
Key off	Sockets 7 and 8		400 to 800 ohms	If any test fails, first check for defective harness to distributor. If harness checks good, distributor stator assembly must be replaced.
	Socket 6 and eng. grd.		0 ohm	
	Socket 7 and eng. grd.		70,000 ohms or more	
	Socket 8 and eng. grd.		70,000 ohms or more	
	Socket 3 and coil tower		7000 to 13,000 ohms	Check coil on coil tester.
	Ignition coil primary terminals		1.0 to 2.0 ohms (Dura II)	Check coil on coil tester.
	Socket 5 and eng. grd.		More than 4.0 ohms	Check for short to ground at DEC terminal of coil or in primary coil winding.

Distributor Hardware Test

1. Disconnect three-wire weatherproof connector at distributor pigtail.
2. Connect a dc voltmeter on 2.5-volt scale to two parallel blades. With the engine cranking, the meter needle should oscillate.
3. Remove distributor cap and check for visual damage or misassembly.
    a. Sintered iron armature (four-, six-, or eight- toothed wheel) must be tight on sleeve, and the roll pin aligning armature to sleeve must be in position.
    b. Sintered iron stator must not be broken.
    c. Armature must rotate when engine is cranked.
4. If the hardware is OK, but the meter doesn't oscillate, replace the magnetic pick-up assembly.

    If above checks are OK, substitute a known good module. If this corrects malfunction, reconnect original module to verify that connector is not causing problems.

TROUBLESHOOTING THE FORD SOLID-STATE 1977 IGNITION SYSTEM (DURA-SPARK SYSTEMS I, II)

Circuit Checks

1. The following voltage and resistance test checks inputs to the module.
2. The module *must* be disconnected from the harness for all tests except the Ignition coil BAT terminal to ground voltage test on the Dura Spark II system. Make all these tests at the *harness* side of the *module* connectors except the one made at the coil tower.

Distributor Hardware Test

1. Disconnect three-wire weatherproof connector at distributor pigtail.
2. Connect a dc voltmeter on 2.5-volt scale to two parallel blades. With the engine cranking, the meter needle should oscillate.
3. Remove distributor cap and check for visual damage or misassembly.
   a. Sintered iron armature (four-, six-, or eight-toothed wheel) must be tight on sleeve, and the roll pin aligning armature to sleeve must be in position.
   b. Sintered iron stator must not be broken.
   c. Armature must rotate when engine is cranked.
4. If the hardware is OK, but the meter doesn't oscillate, replace the magnetic pick-up assembly.

If above checks are OK, substitute a known good module. If this corrects malfunction, reconnect original module to verify that connector is not causing problems.

PRESTOLITE BREAKERLESS INDUCTIVE DISCHARGE (BID) IGNITION SYSTEM

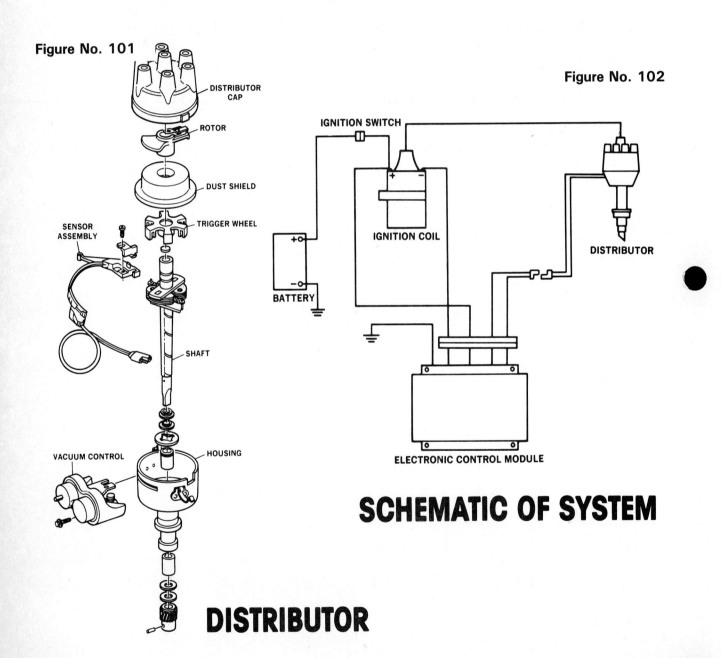

PRESTOLITE BREAKERLESS INDUCTIVE DISCHARGE IGNITION SYSTEM

The Prestolite breakerless inductive discharge igntion (BID) system is a three-part system consisting of a distributor, control unit, and coil. The distributor differs from the conventional design only in that the cam, contact, and condenser assemblies have been replaced by a trigger wheel and sensor. The ignition coil is of conventional design modified to provide maximum system performance.

The operation of the sensor and trigger wheel is such that, when one of the edges of a tooth is aligned with the centerline of the sensor, a trigger signal is sent to the control unit, which in turn opens the coil primary circuit the same as the contact set does in a standard distributor. Opening the primary circuit of the coil develops a high voltage at the tower terminal of the ignition coil. This high voltage is transmitted in the conventional manner to the various spark plugs through the distributor cap and rotor. The dwell time is determined by the space between adjacent teeth and is not adjustable throughout the service life.

The control unit shapes and amplifies the impulse from the sensor so that it may turn on and off a power transistor in the control unit. This power transistor acts much like the contact set in a conventional distributor in that it allows the coil primary to charge to its proper current level, and then suddenly opens the circuit to allow the high-voltage pulse to form at the secondary. An additional feature of the control unit is that a ballast resistor is not used. The primary coil current is regulated electronically. Other features are reverse polarity protection and the ability of the system to be used on either a positive or negative ground electrical system. The control unit is a completely solid state unit which is sealed in a waterproof and vibration-resistant potting compound. The control unit is nonadjustable.

© Copyright 1986, Tune-Up Manufacturers Institute

TROUBLESHOOTING THE PRESTOLITE-AMC BREAKERLESS INDUCTIVE DISCHARGE (BID) IGNITION SYSTEM

Figure No. 103

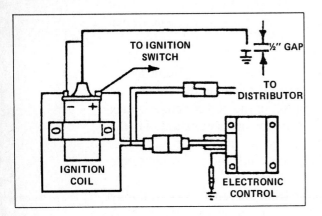

Figure No. 104

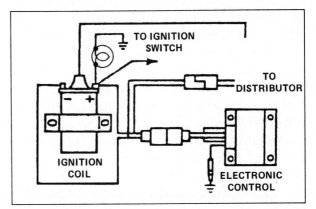

Figure No. 105

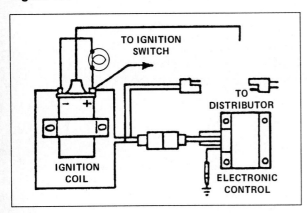

Figure No. 106

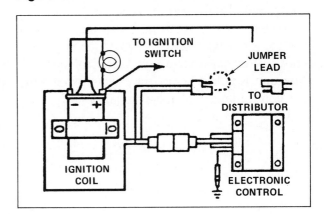

Figure No. 107

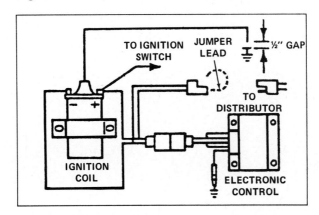

TROUBLESHOOTING THE PRESTOLITE BID IGNITION SYSTEM

The following is a simple six-step troubleshooting sequence for the BID ignition sytem. Equipment needed: one 12-volt test light (No. 57 bulb), and an ohmmeter.

Ignition Switch On

1. Remove coil wire from distributor cap and hold it ½ inch from a good ground. Crank engine and watch for steady sparking from the coil wire. If sparks occur the trouble is not in the ignition system. (Figure No. 103).
2. If no sparks occur (Figure No. 103):
   a. Connect the test light from the coil positive (+) terminal to ground.
   b. Turn the ignition switch to both *on* and *start*.
   c. The bulb should light in both positions. If it *does not,* the fault is in the circuit between the battery and coil.
3. If bulb lights in both positions (Figure No. 104):
   a. Connect the test light across the coil positive (+) and negative (−) terminals.
   b. Unplug the sensor leads near the distributor and turn the ignition on.
   c. If the bulb *does not* light, check the control unit ground connection. If ground is OK, the control unit is defective and must be replaced.
4. If the bulb lights in step 3 (Figure No. 105):
   a. Short across the terminals in the sensor lead coming from the control unit.
   b. If bulb *does not* go off, replace the control unit.
5. If the bulb *does* go off in step 4 (Figure No. 106):
   a. Remove the test light.
   b. Reestablish the ½-inch gap from the coil wire to ground.
   c. Short across the terminals of the sensor lead as in step 4.
   d. If spark *does not* occur each time the terminals are shorted, replace the coil.
6. If a spark occurs whenever the terminals are shorted, the problem lies with the sensor or control unit. To check the sensor (Figure No. 107):
   a. Connect an ohmmeter to the sensor leads at the distributor. Resistance should be 1.6 to 2.0 ohms.
   b. Tap sensor lightly with a pencil; resistance should not vary.
   c. Disconnect one ohmmeter lead from the sensor lead and touch it to the center core of the sensor. The ohmmeter must show an open circuit.
   d. If sensor passes tests a, b, and c, replace the control unit.

Figure No. 108

TYPICAL DUAL-MODE IGNITION SYSTEM

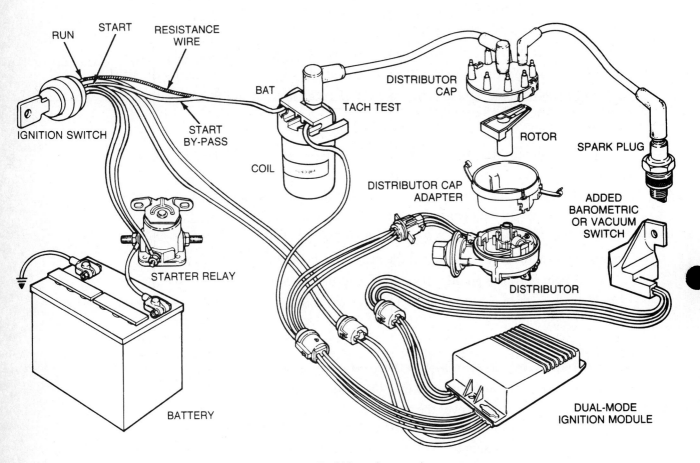

(Courtesy Ford Motor Company.)

FORD DUAL-MODE TIMING-IGNITION SYSTEMS (1978–1980)

To improve high-altitude drivability or low-altitude fuel economy, Ford modified the Duraspark II ignition system. For vehicles operating in high-elevation areas (above 4000 feet), an altitude-compensating device is used to signal the module. Vehicles sold in low-altitude areas employ a manifold vacuum switch. The module used for both altitude compensation and fuel economy is the same and can be identified by its three-plug connectors and two yellow sealing blocks where the wires enter to module case.

Altitude-Compensation System

Since vehicles sold in high-altitude states will sometimes be operated at lower elevations, it is necessary to modify the advanced ignition timing schedule used in low-barometric-pressure environments. A barometric pressure switch is used to signal the module whenever barometric pressure increases to a point indicating an operating elevation of less than 4000 feet. Upon receiving this signal, a circuit in the module retards the timing 3 to 6 degrees. This timing retardation eliminates the detonation problem common to cars calibrated for high-altitude operation.

Economy Calibration System

Operating an engine at greater timing advance increases manifold vacuum and fuel economy. Unfortunately, advanced timing encourages detonation, which could be harmful to the engine and annoying to the driver. A manifold pressure switch signals the module whenever manifold vacuum drops below a specified value. This signal causes the module to retard timing 3 to 6 degrees. During light load situations (cruise), the switch remains closed, allowing the module to maintain advanced spark timing for improved fuel economy. When heavy loads (acceleration) are applied, manifold vacuum drops, causing the switch to open. The opening of the switch creates a voltage drop, signaling the module to retard timing, thus reducing the possibility of detonation.

Figure No. 109

TFI IGNITION SYSTEM

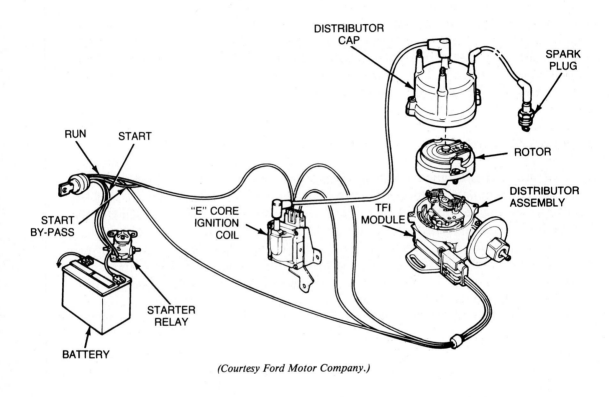

(Courtesy Ford Motor Company.)

THICK-FILM INTEGRATED (TFI) SYSTEMS

Thick film refers to the circuit manufacturing process used in the TFI module. It was introduced on the 1982 model year 1.6-liter engines with automatic transmissions (ATX) and is denoted as the TFI-I system.

Operation of the TFI-I system is the same as Duraspark systems. The differences are limited to construction and location. The small plastic module is attached directly to the distributor, using the distributor housing surface as a heat sink. This installation reduces the size and complexity of the wiring harness.

The major differences lie in the ignition coil. Instead of an oil-filled can type, it is potted in plastic resembling the General Motors remote HEI coil. No ballast resistance is used on the primary side, and internal primary resistance is very low, resulting in greater reserve capacity.

Later TFI-I ignition systems utilize a barometric pressure switch and are similar in operation to the dual-mode timing systems explained previously.

With the introduction of the 1983 electronic fuel injection–electronic engine control IV (EFI-EEC IV) system, the TFI-IV system came into existence. All multipoint fuel-injected models (including turbocharged applications) utilize this system. While the components look the same as those used with TFI-I, the operation is very different. Control of spark advance is performed by the EEC-IV computer (ECA), and no vacuum or centrifugal advance mechanisms are used. Instead of using a reluctance-type signal generator (pick-up coil), the distributor contains a Hall-effect switch. The Hall-effect switch generates a square-wave signal that is more compatible with the digital computer. The Hall-effect switch is referred to as the profile ignition pick-up (PIP) and supplies the computer with crankshaft position and rpm information. This information is combined with other sensory inputs, which permits the computer to send the spark output signal (SPOUT) through the TFI module to control the coil primary circuit.

TFI-IV distributors also use a different rotor. Unlike the conventional blade-type rotor wiper, the TFI-IV rotor has a three-pin type of configuration. These rotors cannot be interchanged. The pin-type rotors also do not require the application of a silicone dielectric compound to reduce radio frequency interference (RFI).

Figure No. 110

HALL-EFFECT SWITCH

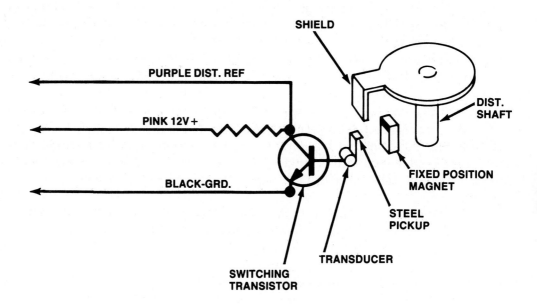

(Courtesy AC-Delco, General Motors.)

HALL-EFFECT SWITCH

The Hall-effect switch is rapidly replacing the traditional ac generators (reluctor/pick-up coil) used on electronic ignition systems of the 1970s. The Hall-effect switch has many advantages over the reluctance type system:

1. It has greater accuracy for determining crankshaft position.
2. It is able to generate a signal at 1 rpm.
3. The digital (square wave) signal is compatible with on-board computers.
4. It can stagger the signal to control timing for individual cylinders, such as needed on odd-fire V-6 engines.

Chrysler began using the Hall-effect switch in 1978 on domestic four-cylinder applications. Ford, General Motors, and Bosch-equipped cars are switching to this system with each new model introduction.

To best understand the operation of Hall-effect switch, do *not* look at it as an ignition system. Instead, consider it a sensory input device for computerized engine control. The Hall-effect switch is responsible for supplying two pieces of information: engine speed (rpm) and crankshaft position (degrees). A computer uses these inputs to determine how much timing advance the engine requires for optimum operating efficiency.

The Hall-effect assembly consists of a fixed-position permanent magnet, shutter blade rotor, pick-up, and a switching transistor. The collector terminal of the switching transistor is connected to a resisted 12-volt source and to the Hall-effect signal wire. The signal wire sends a square-wave message to the computer. The emitter side of the transistor is connected to ground. The base is attached to the pick-up.

When a window of the shutter blade rotor is positioned between the magnet and pick-up, the magnetic field acts on the base circuit of the transistor. This magnetism results in a voltage that turns on the transistor, completing the resisted 12-volt lead to ground. The voltage in the signal line reflects the voltage drop across the resistor and sends approximately 1 volt to the computer.

As a shutter blade passes between the magnet and pick-up, the magnetic field is blocked. Without magnetism striking the pick-up, the base circuit opens, turning off the transistor. The resisted 12-volt line is no longer grounded. Current flow ceases and the resistor drops out of the circuit, resulting in a 12-volt signal to the computer. It is important to note that the shutter blades must be grounded. Should the shutters lose ground, the blades would not be able to block the magnetic field, leaving the transistor turned on all the time.

With a running engine, the shutters and windows of the rotor are alternately passing between the magnet and pick-up. Thus the signal wire voltage is constantly switching between 1 volt and 12 volts. This sharp rise and fall in voltage is referred to as square wave generation or *chopping*. Since late-model computers are digital types, this on–off signal is compatible with the computer circuits. AC-generator-type pick-ups create an analog signal that must be converted to digital form for the computer. The accuracy and simplicity of Hall-effect switches make them a natural choice for future applications.

Figure No. 111 H.E.I. ELECTRONIC SPARK TIMING/ELECTRONIC SPARK CONTROL

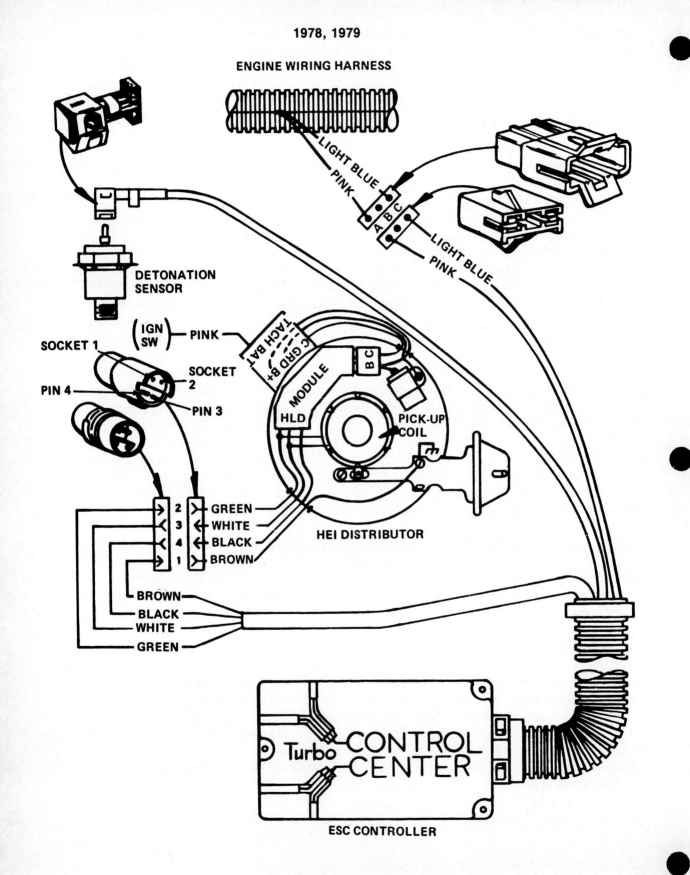

(Courtesy AC-Delco, General Motors.)

DETONATION SENSORS

Detonation has always been a problem for high-compression and turbocharged engines. Their greater combustion chamber pressures cause the fuel charge to self-ignite, resulting in a higher rate of expansion than the cylinder can accommodate. The increased heat and rapid expansion cause engine *knock,* which is why detonation sensors are often called knock sensors.

Since higher-compression engines are more efficient, some method of detonation control is necessary. Water-injection and fuel-enrichment systems were used in the past, but neither approach could be controlled accurately and both interfered with emission calibrations. With the advent of microprocessor and computer controls, it became possible to limit knock electronically with the detonation sensor. The detonation sensor has made it possible to improve engine power and efficiency by allowing higher compression ratios and greater ignition advance. Instead of retarding the ignition advance across the entire timing schedule, the timing is only retarded when knock is detected by the sensor. The detonation sensor first appeared on 1978 General Motors turbocharged vehicles. Today the sensor can be found on virtually any engine, as well as on all turbocharged applications.

When detonation occurs, vibration shock waves are transmitted through the engine. These shock waves strike the sensor and apply pressure to the sensor element. The element is a piezoelectric component, which responds to the shock waves by generating a small ac voltage signal. A piezoelectric crystal generates an electrical voltage when subjected to mechanical stress (pressure). Quartz crystals are the most common piezoelectric devices.

On Chrysler and Ford applications, the detonation sensor is connected directly to the system computer. When knock is sensed by the crystal, the computer is informed of a detonation condition. The computer retards the timing until the sensor no longer sends a signal. Timing advance is restored when a detonation signal is no longer received.

1982-Up General Motors Electronic Spark Control (EST ESC CCC)

Unlike Ford and Chrysler, General Motors does not connect the detonation sensor directly to the electronic control module (ECM). Instead, the sensor is wired to an electronic spark control (ESC) controller, which in turn signals the computer. The ESC controller receives power at terminal F and is grounded through terminal K. Detonation sensor inputs are wired to terminal B and signal outputs are sent to the computer (ECM) from terminal J.

With the engine running at between 1500 and 2000 rpm and *no* knock present, the sensor emits a voltage signal of approximately 0.080 volt ac (80 millivolts ac). Since this signal does not indicate the presence of detonation, the ESC controller sends an 8- to 10-volt reference to the ECM. The computer is able to advance timing according to the electronic spark timing (EST) program.

Should knock occur, the detonation sensor output increases, signaling the controller that the ignition timing must be retarded. The controller removes the 8- to 10-volt reference signal from the output line to the ECM. The ECM is informed that detonation is present and retards the timing until the controller resumes the 8- to 10-

volt signal. Should the voltage from terminal J be removed for prolonged periods of time, the computer sets a code 43.

Testing Detonation Sensors

While the detonation sensor may seem complicated, it is a very simple device to test. The following procedure can be used for most detonation-sensor-equipped cars:

1. Raise the engine speed to approximately 2000 rpm. Placing the throttle on the high step of the fast idle cam is usually the best method to use for holding higher engine speed.
2. If the detonation sensor is located on the intake manifold, tap repeatedly on the manifold in the vicinity of the sensor. Engine-block-mounted sensors are more difficult to reach. In this case, tap on the exhaust manifold with a plastic hammer.
3. While tapping, the engine speed should drop as the timing is retarded. No change in speed indicates a defective detonation sensor.

It should also be noted that defective or incorrectly routed ignition wires can affect the knock sensor signal. Such a condition would test out as a defective sensor.

Servicing Detonation Sensors

The installation of a knock sensor is no more difficult than changing a spark plug. Because it is a finely calibrated device, certain precautions must be observed:

1. Start sensor threads by hand. Be very careful not to "cross" the threads. If the manifold threads are damaged, the manifold must be replaced.
2. Torque sensor to manufacturer's specifications. This does not normally exceed 14 pound-foot. Overtorquing alters the sensor calibration.
3. Reconnect the sensor lead and test system for proper operation.

NOTE: Do not apply side force to the sensor when installing.

Notes

Figure No. 112

EST SCHEMATIC

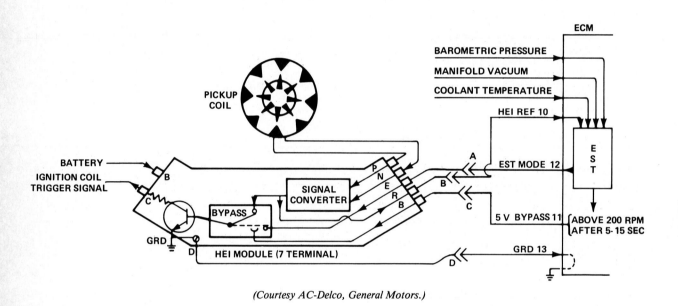

(Courtesy AC-Delco, General Motors.)

GENERAL MOTORS ELECTRONIC SPARK TIMING (EST CCC 1980-ON)

During the latter half of the 1970s, General Motors developed several types of microprocessor-controlled ignition systems. Such systems are electronic spark selection (ESS), electronic module retard (EMR), and HEI MIZAR were used to better control ignition timing for improved drivability and reduced emissions. These systems were found on higher-priced vehicles and limited to only a two- to three-year production run. The advent of computer command control (CCC) in 1980 brought the development of the electronic spark timing system (EST CCC). When compared to the previous attempts at microprocessor control, the EST CCC is more accurate, less complicated, and free of the intermittent problems common to those early systems. The EST CCC is a totally electronically controlled system. No centrifugal or vacuum-advance mechanisms are used to control ignition timing advance. The electronic control module (ECM) is the computer center for the system and varies the ignition timing in relation to the information received from the following sensors:

1. Manifold vacuum
2. Barometric pressure
3. Coolant temperature
4. Throttle position
5. Engine speed and crankshaft position

The ability of the ECM to monitor all these variables makes it possible to more accurately control ignition timing. This greater accuracy leads to vastly improved performance and reduced emissions.

The Ignition Module

Instead of using the three-, four-, or five-terminal (pin) HEI module, the EST CCC distributor contains a seven-terminal HEI module. The integral coil distributors (135 mm) and the remote coil distributors (89 mm) contain modules that are different in external size and shape, but incorporate the same internal circuitry. Internal circuitry changed with the 1982 model year, but operation remained the same.

To best comprehend the operation of the EST CCC system, it is important to understand the function of each individual terminal on the module.

The coil side of the module consists of the *B+ and C terminals*. The B+ terminal is connected to the positive (+) side of the coil, which supplies available voltage to the ignition circuit. The C terminal is responsible for controlling the flow of current through the coil primary circuit. As is the case with any coil-fired ignition system, the primary circuit is opened and closed to ground for secondary discharge control. A computer-controlled system controls the coil primary in the same manner as a conventional or electronic ignition system.

The control side of the module consists of the P-N-E-R-B terminals. The *P and N terminals* connect the module to the pick-up coil and are the same as those terminals used on the standard four-terminal HEI module. The ac wave signal supplied by the pick-up coil is converted to a digital (on–off) pattern by the signal-converter section of the module. The converted ac signal is used for both controlling the coil primary and signaling the ECM, depending on the operating mode.

The *E terminal* is the EST wire. When the ECM assumes control of the ignition system, it will use the E terminal as the computer command signal line. The ECM controls the coil primary circuit current flow through the EST wire.

The *R terminal* is the HEI reference signal to the ECM. This terminal acts as a sensory input for the computer by supplying engine speed (rpm) and crankshaft position information. If the ECM does not receive a reference signal, it cannot recognize that the engine is running. By counting the time between digital (on–off) signals, the ECM calibrates the crankshaft position.

The *B terminal* is a 5-volt bypass output signal connection from the computer. When the ECM applies 5 volts to the B terminal, the HEI module switches coil primary control to the EST wire. If no voltage is present, the module controls coil primary current flow.

EST CCC Operation

During cranking, and for the first 5 to 15 seconds of engine operation, the module controls the primary current flow in the same manner as the old four-terminal HEI module system. This period of operation is referred to as the *module mode*. While in module mode, timing advance is limited to approximately 5 degrees. Should the engine remain in this mode, performance would be severely limited due to the lack of sufficient timing advance.

By monitoring the HEI reference signal, the ECM can recognize that the engine is running. When a predetermined rpm and elapsed run time have been achieved, the ECM applies 5 volts to the B terminal of the HEI module. This voltage causes the module to connect the EST line to the base of the power transistor and allows computer control of the coil primary. The engine now runs in the *EST mode*. The EST mode permits the ECM to control timing. The pick-up coil is used only for HEI reference-signal generation, while in this mode.

Diagnostics

The ECM contains circuitry that enables the system to self-diagnose problems. Problems in the system are displayed in the form of diagnostic trouble codes, which are explained in the computerized engine control section of the text.

Code 12 indicates that there is a loss of the HEI reference signal to the computer. Since the ECM cannot determine engine speed, no 5-volt bypass signal is sent to the module. The ignition system would remain in the module mode with timing advance restricted to 5 degrees. Should the reference line go open during engine operation, the check engine light would come on, but code 12 is not stored in the diagnostic memory.

In 1982, code 41 was added to the diagnostic memory. Code 41 indicates an open in the HEI reference line, but differs from code 12. When the ECM sees a manifold vacuum signal that indicates a running engine and does not receive the HEI reference, the check engine light comes on and code 41 is stored in the memory.

Code 42 reveals a problem in the EST line. An open EST wire is recognized by the ECM as an abnormal condition and sets a code 42. If this condition were to occur, the engine would stop, but restart and run in the module mode. A grounded EST line also sets a code 42. If the EST line should become grounded while the engine is running, the engine would stop and not restart. Anytime the engine does not start and a code 42 is in the memory, look for a grounded EST line.

Notes

8
FUEL SYSTEMS

FUEL SYSTEMS

Figure No. 113

MECHANICAL PUMP—ENGINE MOUNT

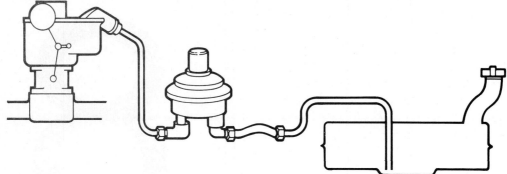

Figure No. 114

IN-LINE ELECTRICAL PUMP

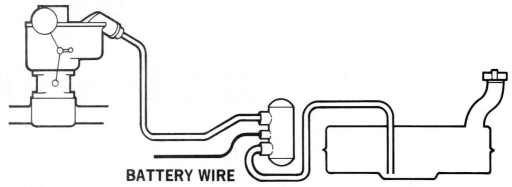

Figure No. 115

IN-TANK ELECTRICAL PUMP

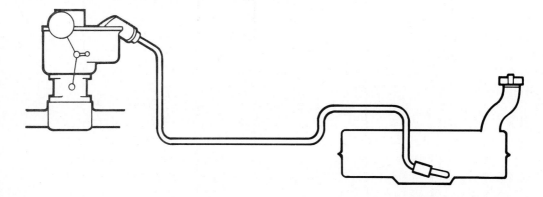

© Copyright 1986, Tune-Up Manufacturers Institute

FUEL SYSTEMS

Carburetor-Equipped Internal Combustion Engines

The fuel system has several jobs to do. The first is to store fuel so that it is available for use when needed. Another job is to deliver this fuel to the carburetor. Still another is to mix the fuel with the proper amount of air regardless of speed or load conditions and to control the amount of fuel and air entering the engine to meet the desires of the driver and the needs of the engine. The components of the fuel system are fuel tank, fuel line, flex line, fuel pump, fuel filter, carburetor, air cleaner, and intake manifold

The fuel tank is a reservoir built into the vehicle to contain a reasonable supply of gasoline for the fuel system. It must be gasoline tight so that no fuel is lost, and it also must have an air vent either in the tank cap or through the fuel evaporation emission control system. Venting is extremely important since the lack of an air vent or a plugged air vent will very often cause what appears to be a fuel-pump failure. Most gasoline tanks are constructed so that the gasoline pick-up is a short distance above the bottom of the tank. This permits rust, water, and other foreign matter to collect on the bottom of the tank but not be pulled into the gas line and through the fuel pump.

Mechanical Electrical In-Line

Mechanical and electrical in-line fuel-pump installations consists of three basic parts: the pipe or tube from the gas tank to the flexible hose, the flexible hose from the gasoline pipe to the fuel pump, and the pipe from the fuel pump to the carburetor. The purpose of the flexible hose is to prevent damage to the gas line from engine vibration. Engines are mounted on resilient rubber engine mounts and, consequently, the vibration between the engine and the frame would soon cause a break in a metal gasoline line.

In operation, the fuel lines from tank to hose and from hose to pump are subjected to vacuum or pressure below atmospheric pressure. Consequently, leaks in either of these will result in air entering the gasoline before it reaches the fuel pump. Another point to consider is that the flexible hose may collapse and partially or wholly restrict the flow of fuel from the tank to the pump. The fuel line from the pump to the carburetor is under constant pump pressure.

The mechanical and electrical in-line pumps draw gasoline from the fuel tank and pump it up into the carburetor fuel bowl. In-line electric pumps characteristically develop less vacuum and normally are located close to the tank. The in-tank electric pump is submerged in the tank, and fuel is pushed from the tank to the carburetor. The pump must supply a sufficient quantity of fuel, at the correct pressure, to the carburetor at all times to meet the engine fuel requirements under all operating conditions.

The function of the carburetor is to mix the right amount of fuel with the right amount of air and to deliver this carefully compounded air-fuel mixture to the engine under all conditions of speed and load.

Gasoline is a mixture of a number of compounds called hydrocarbons. It is composed of about 15 percent hydrogen and about 85 percent carbon. These substances unite with the oxygen in the air at the time combustion occurs, changing the hydrogen and carbon primarily into water, carbon monoxide, and carbon dioxide. (Further discussion of this subject is covered in Chapter 9.) The burning of gasoline generates high pressures in the combustion chambers, which is exerted on the heads of the pistons, thereby powering the engine.

Gasoline has a potential energy three times greater than TNT and considerably greater than dynamite. Gasoline burns only when it is exposed to air. If an open can is filled with gasoline and the gasoline is lighted, there will be no explosion. The surface of the gasoline will merely burn and the heat in the gasoline will be used up very slowly. For an explosion to occur with the gasoline, and an explosion is really a rapid burning process, more surface of the gasoline must be exposed to the air. This is why the carburetor is designed to spray or atomize the gasoline into the air stream so that the heat energy contained in the fuel can be more completely liberated for maximum power development.

Notes

Figure No. 116

MECHANICAL FUEL PUMP

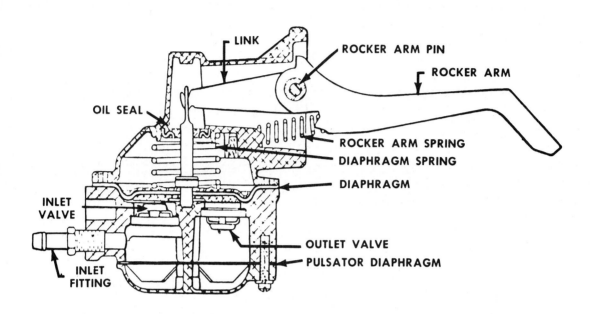

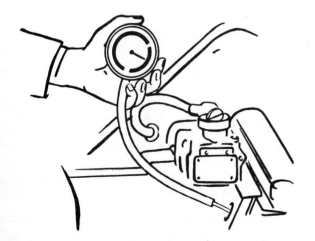

Pressure Test

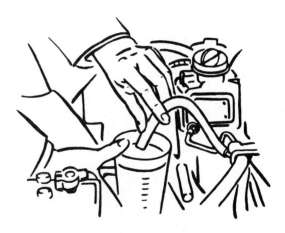

Volume Test

© Copyright 1986, Tune-Up Manufacturers Institute

FUEL PUMP

The mechanical fuel pump is operated by the action of an eccentric on the camshaft, which actuates the pump rocker arm. The rocker arm spring holds the arm in constant contact with the camshaft eccentric.

As the pump rocker arm is moved downward by the action of the eccentric, it bears against a shoulder on the link which pivots on the rocker arm pin. The link swings upward, pulling the diaphragm upward and compressing the diaphragm spring. The action of the rising diaphragm creates a vacuum in the fuel chamber located under the diaphragm. This action pulls the outlet valve closed and causes fuel from the gasoline tank, which is under atmospheric pressure, to enter the pump fuel chamber, which is under reduced pressure, through the inlet valve.

Further rotation of the camshaft eccentric permits the rocker arm to swing upward due to the action of the rocker arm spring. The arm releases the diaphragm link. It cannot force the link downward because it works in an elongated slot in the link.

The compressed diaphragm spring then exerts pressure on the diaphragm, which in turn applies pressure to the fuel in the chamber below. The maximum pressure exerted on the fuel depends entirely upon the strength of the diaphragm spring. The pressure on the fuel forces the inlet valve closed while forcing the fuel out of the outlet valve into the carburetor.

Pump Action

Pump action delivers fuel to the carburetor only when the pressure in the pump outlet line is less than the pressure exerted by the diaphragm spring. When the demand for fuel is low, and the carburetor float chamber is filled, the carburetor float presses the needle valve on its seat, shutting off the entrance of more fuel from the pump.

At this time the pump builds up pressure in the fuel chamber until it overcomes the pressure of the diaphragm spring. This results in almost complete stoppage of further movement of the diaphragm until more fuel is needed.

The pulsator diaphragm in the bottom of the pump provides a form of air pocket which is compressed by the pressure on the fuel. When the pump is on the suction stroke and the pressure in the pulsator chamber is relieved, the compressed air will push the fuel out of the outlet valve as soon as the carburetor needs more fuel. The pulsator diaphragm also minimizes the pulse surges that are experienced by normal pump action.

A fuel pump is tested for pressure and volume. A typical example of a test specification is:

      Pressure: 5¼ to 6½ pounds at idle rpm and at 1000 rpm
      Volume: 1 pint in 30 seconds or less at idle rpm

While conducting fuel pump tests, *observe all fire precautions.*

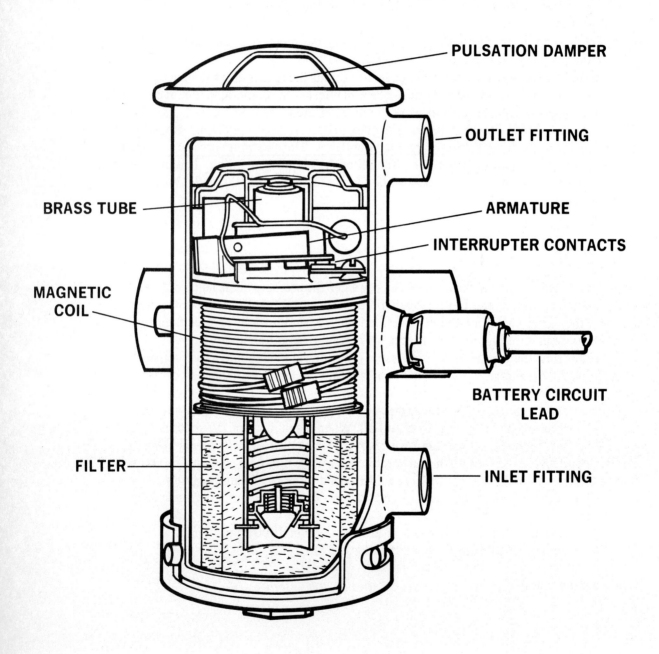

ELECTRIC FUEL PUMPS

Electric fuel pumps fall into two different categories or types, *positive displacement* and *centrifugal.*

The two most common types of positive-displacement pumps in the automotive field are the *reciprocating* and *rotary-vane* designs. The reciprocating pump is a low-pressure unit and is mounted in the fuel line. The rotary-vane pump can be a low- or high-pressure unit and can be mounted either in the line or in the tank. The centrifugal pump is also a low-pressure unit and is mounted in the tank; it is referred to as a pusher-type pump.

Reciprocating Pump

There are various electric pump designs with this category. Pumping action is accomplished by means of a diaphragm, bellows, or plunger. However, the basic pump operation is similar within this group. The term reciprocating or oscillating refers to the back and forth motion of the plunger, diaphragm, or bellows.

Operation

The unit in Figure No. 117 is a plunger type, solenoid-activated fuel pump. Pumping action results from the reciprocating motion of the plunger. When the ignition is turned on, battery current flows through the closed interrupter contacts, energizing the magnetic coil. The plunger is forced downward by the strong magnetic field, compressing the plunger spring. A permanent magnet, located on the interrupter armature, senses the position of the plunger and opens the contact set, resulting in the collapse of the magnetic field. The compressed plunger spring then forces the plunger in the opposite direction, creating a vacuum in the lower cavity below the plunger. Fuel, which is under atmospheric pressure in the tank, flows into the pump's inlet fitting and into the lower pump cavity.

From there, it passes through the filter and up through the first inlet check valve, filling the brass tube below the plunger. The contacts close and the plunger again is forced downward, closing the inlet valve. A second check valve opens in the head of the plunger allowing the fuel in the lower cavity to enter the upper cavity of the pump via the brass tube. Again the permanent magnet senses the position of the plunger, opening the contacts, causing a collapse of the magnetic field. The plunger spring pushes the plunger upward, closing the plunger check valve, forcing the fuel out of the upper cavity through the outlet fitting and to the carburetor.

Frequency of the plunger oscillation depends on the fuel flow. When operating at maxiumum delivery and zero pressure, the pump oscillates with a frequency of approximately 25 cycles per second; when operating at maximum pressure and zero delivery, plunger oscillation is approximately one or two cycles per second. Fuel delivery pressure is developed by the force of the return spring acting on the plunger area. Depending on the unit design, static pressure can range from 1¼ to 7½ psi and free flow deliveries range from 18 to 45 gallons per hour. Standard operating voltages are 6, 12, and 24 volts dc.

The interrupter and magnetic coil cavity are sealed off from the fuel being pumped. This cavity is filled with an inert gas to prolong contact life. The upper pump cavity is closed off by means of a spunover steel cap which also holds down a pulsation damping diaphragm and enclosing an air cavity. Thus, delivery of fuel from the pump is relatively smooth without sharp pulsations.

Notes

FLOAT SYSTEM

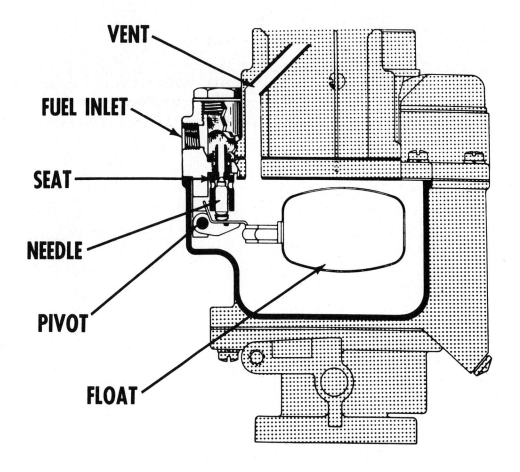

FLOAT SYSTEM

The automobile is driven at variable speeds under a variety of operating conditions. Each condition has its own specific fuel requirement. The carburetor is designed with several different systems or circuits, each with its own particular function, to supply these varying needs. These systems or circuits are the float system, idle circuit, main metering system, power system, accelerating system, and the choke system. Each system will be discussed in turn.

The float system maintains a constant supply of fuel in the float chamber for immediate use by the fuel-metering systems in the carburetor. Fuel, under pressure from the fuel pump, enters the float chamber through the fuel inlet line. This fuel raises the float on its pivot, thereby controlling the needle valve and admitting only enough fuel to replace that being used. As the fuel level drops due to the fuel being consumed by engine operation, the float lowers, permitting fuel-pump pressure to unseat the needle valve and admit more fuel to the bowl. The entrance of fuel again raises the float. When the correct level is reached, the float again closes the needle valve on its seat, shutting off the fuel supply from the pump. Actually, the needle valve will drop only slightly as the fuel is consumed, since the action of the valve is quite sensitive.

The level at which the float maintains the fuel supply in the float chamber is extremely important. If the correct fuel level is 3/16 inch below the tip of the main discharge nozzle, an increase of 1/32 inch would decrease the distance that the fuel has to be lifted by 16 percent, resulting in a fuel mixture that is too rich. A decrease of 1/32 inch in fuel level would increase the distance that the fuel has to be lifted by 16 percent, resulting in a fuel mixture that is too lean.

The float chamber is vented to the atmosphere through a port in the air horn. The function of the vent is to allow the fuel to be smoothly withdrawn from the chamber into the various carburetor circuits.

IDLE CIRCUIT

An Idle Circuit is needed that will deliver some gasoline and air when the throttle is closed

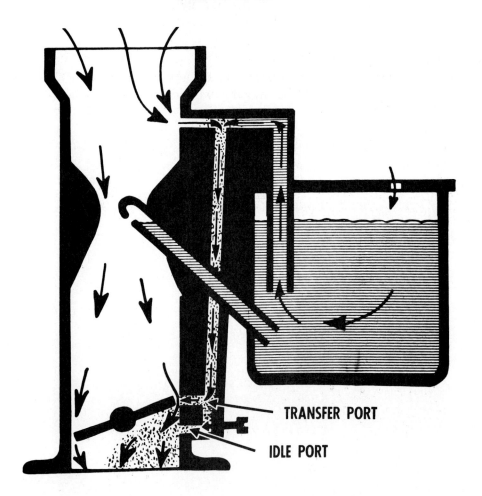

TRANSFER PORT
IDLE PORT

IDLE CIRCUIT

When the throttle valve is closed, there is not enough air flow through the carburetor to create a low-pressure area at the main discharge nozzle. Therefore, an idling system has been provided. The idling system delivers fuel through the idle discharge port located below the throttle valve. A drop in pressure is created at this point because the throttle valve acts as a pressure-dropping device when in the closed position. All the air that passes through the carburetor throat must pass around the throttle valve. The fuel is forced through this idle system because of the pressure difference between the atmospheric pressure on the fuel in the fuel bowl and the low-pressure area created by the nearly closed position of the throttle valve.

Fuel drawn from the float bowl is metered by an idle jet and mixed with a controlled quantity of air from one or more calibrated air bleeds. The air-fuel mixture is then discharged from the idle port just below the slightly open throttle, mixing with air passing through the carburetor to form the idle mixture. An adjusting needle at the idle port can be turned to provide a leaner or richer air-fuel ratio.

As the throttle is opened, additional air flows through the carburetor. Although still insufficient to draw fuel from the main nozzle, this increased air flow will result in an excessively lean mixture and will cause a flat spot in engine performance unless additional fuel is made available. This problem is overcome by adding a transfer port just above the closed throttle position. When the throttle is opened, it gradually exposes the transfer port to intake manifold vacuum-causing discharge of air-fuel mixture from this port as well as from the idle port.

Both the idle and transfer ports are carefully designed to provide a smooth transition between idling and cruising speeds of 20 to 40 mph, depending upon carburetor design. At these speeds, the throttle opening and air flow are sufficient to permit the main metering system to supply the necessary fuel. The idle and low-speed system now cease to function due to insufficient vacuum acting on the idle system ports.

Complete carburetor adjustment is covered in the tune-up procedure section of this course.

Figure No. 120

MAIN METERING SYSTEM

Carburetors are designed with a venturi tube positioned in the main carburetor body. The function of this tube is to control the fuel discharge from the main jet by restricting the air opening through the carburetor.

The air taken into the carburetor air horn must pass through the venturi, and in passing through this restricted area the air momentarily increases in speed. This increase in air speed causes a partial vacuum or low-pressure area at the narrow point of the venturi. This low-pressure area is utilized to cause the fuel to flow because of the differential pressure between atmospheric in the fuel bowl and lower than atmospheric in the venturi. As the throttle is opened, the speed of the air flow through the carburetor is increased, with a proportional pressure drop in the venturi.

Since maximum velocity and pressure drop are obtained at the smallest cross section of the venturi, this is the logical point to locate a discharge nozzle for maximum fuel flow. Air entering the carburetor passes through the venturi, where its velocity is increased and its pressure is reduced. Atmospheric pressure in the bowl forces fuel through the nozzle into the air stream. To prevent fuel flow from the nozzle when the engine is not operating, the fuel level in the bowl is maintained slightly below the discharge nozzle opening.

The amount of fuel metered into the air stream is controlled by a main jet or metering orifice. The metering orifice is calibrated in size to provide accurate metering of the fuel in the economy range. In most cases this metering orifice is not large enough to carry a sufficient amount of fuel to supply the engine at high speeds and full load. The metering rod type of system combines both the economy and power ranges. To accomplish this, a jet sufficient in size to supply the required amount of fuel under full load operation is used. To reduce the effective size of the orifice for part throttle operation, a calibrated rod called a *metering rod* is inserted in the jet. This rod is mechanically connected to the throttle shaft and is raised as the throttle is opened.

To provide the proper opening in the orifice at any given throttle position, the metering rod is graduated in various diameters along its lower end, that portion which moves in the orifice. At idle or low-speed operation, the metering rod is inserted in the jet so that the large diameter of the rod is in a position to permit only a minimum of fuel flow through the orifice. As the throttle is opened, the metering rod is raised in the jet, progressively increasing the effective orifice opening and thereby allowing a greater flow of fuel. At wide open throttle, the metering rod is raised to its limit in the orifice, allowing maximum fuel flow for high-speed and full-load operation.

Figure No. 121

VACUUM CONTROLLED METERING ROD

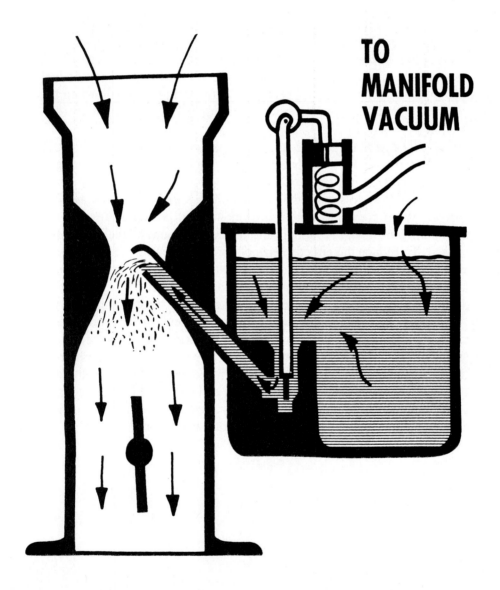

VACUUM-CONTROLLED METERING ROD

The vacuum controlled metering rod provides more metering rod movement with less throttle action than the manually controlled type of metering rod.

When the throttle is closed, engine manifold vacuum is high. Atmospheric pressure pushes the piston down, fully compressing the spring. The metering rod provides maxiumum restriction in the main jet, and very little gasoline flows through the high-speed nozzle.

As the throttle is opened slightly, there is a corresponding decrease in manifold vacuum. The spring pressure then raises the piston and metering rod slightly, permitting an increased fuel flow out of the high-speed nozzle.

When the throttle is opened wide, manifold vacuum drops sharply. The spring pressure pushes the piston and metering rod upward to its full limit of travel. This permits a maximum flow of gasoline through the high-speed nozzle to supply the demand for full power.

During closed or part-throttle operation, the high engine manifold vacuum keeps a larger portion of the metering rod inserted in the main jet, as previously explained. On quick throttle opening, the sudden drop in manifold vacuum permits the spring to push the piston and metering rod upward very quickly. This action permits increased fuel flow to supply the sudden demand for increased power.

Every performance demand made on the engine by the action of the throttle is immediately compensated for by the sensitive action of the vacuum-controlled metering rod.

Figure No. 122

POWER SYSTEM OPERATION

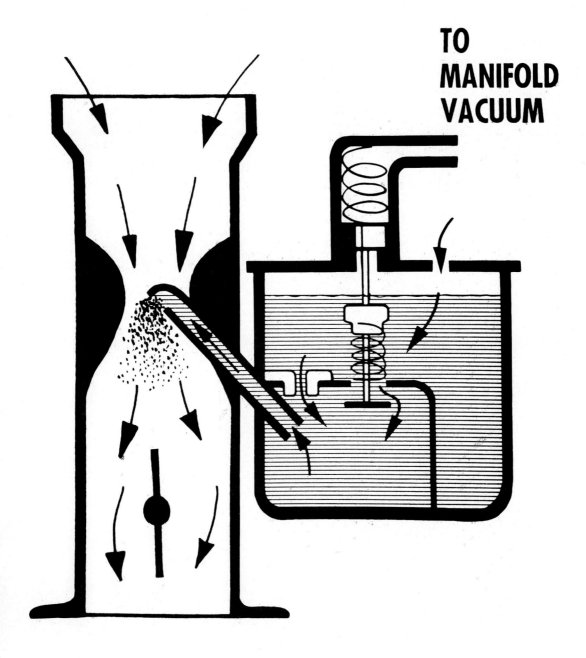

TO MANIFOLD VACUUM

POWER SYSTEM

For normal cruising speeds, an economical mixture of about 16 parts of air to 1 part of fuel is calibrated by the main metering system. For high-speed or full-throttle operation, a richer mixture is necessary. This mixture must be approximately 12 parts of air to 1 part of fuel. The power system automatically supplies the added fuel when it is required by the engine.

The power system may be said to serve two functions. During normal cruising speeds the power system does not operate and may be called the economizer system. Under the demands of full-throttle operation, the system acts as a fuel-enriching device or power system to satisfy the demands for increased power.

The power system has two important parts: a piston, which is a part of the air system, and a valve, which is a part of the fuel system. The power system is actuated by the manifold vacuum, which reflects a true indication of the power requirements of the engine. The vacuum is taken from a point located below the throttle plate and is transmitted through a passage to the top of the power piston. The bottom of the piston is exposed to atmospheric pressure of the air in the bowl.

During idle and low-speed operation, manifold vacuum creates a low-pressure area above the piston, causing the piston to be forced upward by the atmospheric pressure in the fuel bowl. This action closes the power jet and compresses the calibrated spring above the piston. The power system is not operating at this time.

When high-speed or load requirements are placed on the engine, a greater opening of the throttle plate is necessary to maintain speed. When the throttle plate is opened, less resistance is offered to the engine suction, and the manifold vacuum is reduced. The drop in vacuum permits the piston spring to push the piston down, thereby opening the power jet and permitting the entrance of the additional fuel necessary to meet the engine's demands.

The power valve is located in the bottom of the float chamber and is covered by fuel at all times. Fuel from the float chamber flows through the power valve, which acts as a supplementary main jet. The amount of fuel allowed to flow through the power system is metered by restrictions. This restriction prevents the power system from delivering too much fuel at high speeds.

THE ACCELERATING SYSTEM
puts a charge of gasoline into this air space until the gasoline starts from the nozzle.

Figure No. 123

ACCELERATING SYSTEM

When the throttle is suddenly opened, a volume of uncharged air tends to enter the engine. This condition is caused by the fact that the fuel is heavier than the air and is consequently harder to quickly set into motion. Besides, the fuel has a tendency to cling to the wall of the main well, resisting a sudden increase in movement. A lack of fuel exists, therefore, when transferring quickly from the idle system to the main system.

To enable the engine to accelerate rapidly without hesitation, an accelerating pump temporarily supplies the extra fuel necessary for more power until the main metering system can catch up with the needs of the engine. If it were not for the accelerating pump, a large volume of uncharged air would enter the engine, and the engine would "flat spot" or slow down momentarily.

The accelerating pump is linked to the throttle lever. The pump is so constructed that, when the throttle is moved to the closed position, the piston in the pump chamber moves upward, drawing a quantity of fuel from the fuel bowl past a check valve and into the pump chamber. When the throttle is opened, the piston moves downward, forcing the fuel out of the pump chamber into the carburetor venturi where it mixes with the uncharged air. This enriched-mixture discharge covers the time lapse required for the main metering system to come into full operation.

The discharge of the fuel occurs instantly when the throttle is opened. However, a slot in the piston stem allows an overriding action of the linkage. This arrangement subjects the piston to spring tensions, which provides a prolonged discharge of the fuel rather than just a single spurt.

The accelerating system uses a simple pump employing check valves to permit fuel to be taken in and discharged in accordance with the movement of the pump piston. As the piston moves upward in the cylinder, the inlet check ball is lifted off its seat, allowing fuel to be drawn into the chamber from the fuel bowl. The discharge valve seats itself during the chamber loading operation, thus preventing air from the venturi from being drawn in through the discharge nozzle.

As the throttle is opened, the pump piston is forced downward, and the pressure on the fuel causes the inlet check ball to seat itself, thus preventing the fuel from returning to the fuel bowl. The fuel also moves through the discharge passage, lifting the pump discharge valve off its seat and allowing the fuel to move out of the discharge nozzle into the carburetor throat.

Figure No. 124

CHOKE ACTION

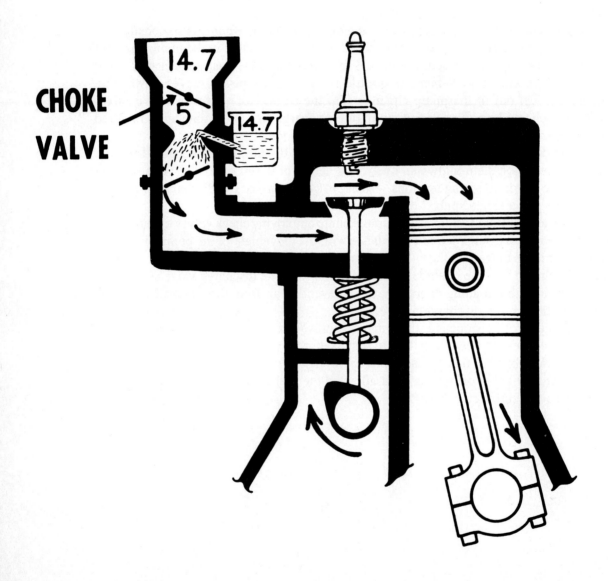

CHOKE

When the engine is cold, a richer mixture than the idle system can supply is required for starting. A choke valve is used to accomplish this. With the choke valve closed, atmospheric pressure is present above the choke valve. As the engine is turned over by the starter, a low-pressure area is created just below the choke valve. As an example, Figure No. 124 shows a pressure of only 5 pounds (equal to 10 inches of vacuum) below the choke valve. This permits air pressure of 14.7 pounds per square inch to push gasoline out of the main nozzle, even though very little air is flowing past this main nozzle. This provides a rich mixture for starting the engine.

A rich mixture is required for starting because with a cold engine atomization of the gasoline is not very efficient, and much of the gasoline condenses on the walls of the intake manifold. Consequently, all the gasoline coming out of the main nozzle is not carried into the cylinders with the air.

The choke valve is mounted on a shaft which is off center. This is to provide a self-opening action of the choke as the engine starts and more air rushes in to meet the needs of the engine. Most chokes are spring loaded or held closed by spring tension. When the engine starts, the choke valve can open sufficiently so that the mixture is not overrich, and the engine will continue to run. With the spring-loaded choke valve, if the throttle valve is closed and less air flows into the engine, the spring tends to close the choke valve, again maintaining about the same pressure differential above and below the choke valve, even though the air volume past the choke varies with engine speed.

Figure No. 125

AUTOMATIC CHOKE

PISTON TYPE

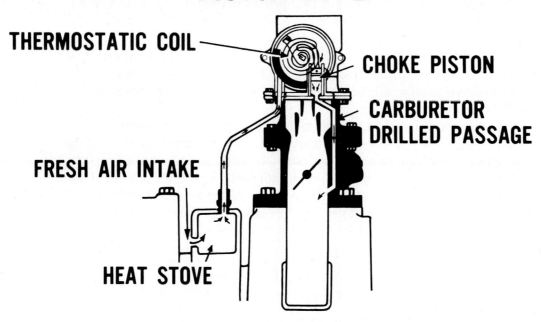

DIAPHRAGM TYPE

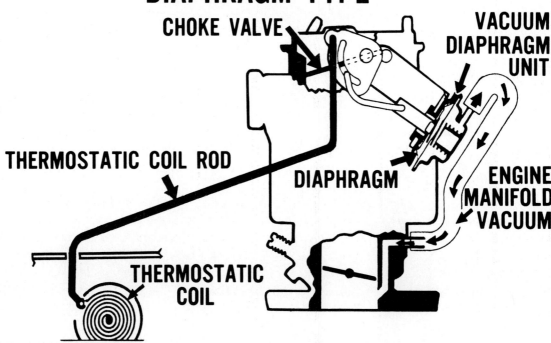

AUTOMATIC CHOKE

Piston Type

The automatic choke is controlled by a combination of a thermostatic bimetal coil spring and a vacuum-operated piston. Both the spring and piston are linked to the choke valve. During cold starting the coil of the spring holds the choke closed. As soon as the engine starts, intake manifold vacuum acting on the piston opposes the spring action, thereby tending to open the choke. As the engine warms up, heated fresh air drawn from the exhaust manifold through a tube causes the thermostatic spring to relax, permitting the choke valve to fully open. Failure of the choke to open properly will cause an excessively rich mixture, fouling of the spark plugs, and poor engine performance.

To counter the opening action of a choke by air velocity, a thermostatic spring tends to hold the choke valve closed. This spring is usually constructed of a strip of brass and a strip of steel fused together and wound into a coil. Since these materials have a different coefficient of expansion, the spring will exert a greater tension when cold than when hot. When the engine is cold, the spring exerts a force on the choke valve to hold it in a closed position. As the engine warms up, this tension gradually decreases in proportion to the increase in engine temperature. Therefore, when the engine is warm, spring tension is relaxed, permitting the choke valve to open.

The vacuum piston is connected by linkage to the choke shaft and operates in a cylinder which is connected by a passage to a point below the throttle valve. This makes the piston subject on one side to the low pressure of the intake manifold. When the choke is moved to the closed position by the thermostatic spring, the piston is moved to one end of the cylinder in which it operates. Vacuum or low pressure is applied to the piston the instant the engine starts. The piston moves toward the low pressure and partially opens the choke. As the piston moves in the cylinder, it encounters slots cut in the side of the cylinder. The piston will only move toward the low-pressure area to a point where the slots provide a passage around the piston. The low pressure is then satisfied by air moving from the atmospheric pressure at the heat stove end of the pickup, through the choke housing, which includes the thermostatic spring, and then around the piston by way of the passage formed by the slots in the piston cylinder. The result is that the choke is allowed to assume a *starting* position as determined by the thermostatic spring, and at the instant the engine starts the vacuum-operated piston provides a *running* position of the choke by opening the choke to a definite minimum position. The piston linkage being connected to the choke shaft results in the vacuum piston canceling part of the tension of the thermostatic spring. As the thermostatic spring continues to uncoil, it simply moves the piston down in its cylinder. Vacuum during this period is bypassed through the cylinder slots.

A choke lock, which is part of the fast idle and choke linkage, is used to keep the choke in the off position should the throttle be held open for an extended time. This will prevent the choke from closing due to the loss of warm air flow through the choke housing caused by the drop in engine vacuum.

A device, called the *choke unloader,* is connected to the throttle lever and choke linkage. This mechanism permits the choke to be opened sufficiently, by flooring the accelerator, to clear the excess gasoline from the engine should the engine flood during starting.

Automatic chokes of this type are adjusted by tensioning the bimetal spring. This is effected by loosening the cover retaining screws and turning the cover to align the emboss mark on the cover with the proper index mark on the choke body, and then tightening the retaining screws. Specifications for an automatic choke setting are, for example, index mark, one notch rich or two notches lean.

As previously stated, air from outside the engine is heated as it passes through the heat stove on its way to the automatic choke. In the event the hot exhaust gases burn out the heat stove, exhaust gases will flow into the automatic choke, presently resulting in carbonizing of the choke passages and mechanism. When the condition is detected, the choke mechanism must be cleaned and adjusted, and the heat stove replaced.

Diaphragm Type (Vacuum Break or Choke Pull-Off)

Some late-model carburetors, such as the Rochester VV, 2GV and 4MV, the Stromberg WWC3, the Carter AVS and YF models, and Holley 2209, 4150 and 4160 models use a diaphragm-type automatic choke. In this choke design, the vacuum diaphragm replaces the choke piston, and the thermostatic coil spring is mounted in the intake or exhaust manifold, over the heat crossover passage.

Engine vacuum below the throttle plate is used, in conjunction with the thermostatic spring, to operate the diaphragm choke in much the same manner as in the piston-type choke.

The adjustment of the diaphragm-type choke is performed by disconnecting the choke rod at its upper end from the carburetor choke lever. Hold the choke valve closed and lift the choke rod upward until it strikes its stop in the thermostat housing. Then check the position of the top of the rod in relation to its hole in the choke lever. The average setting is from ½ to 1 rod diameter above the top of the lever hole. If an adjustment is necessary, bend the choke rod, as required, in its offset area to effect the desired adjustment. Connect the choke rod to the choke lever and check for free choke operation and for complete choke closing.

A choke setting equal to "2 notches" rich in the piston-type choke would be basically 2 rod end diameters above the hole in the choke rod.

Some automatic chokes use the heat of the engine coolant instead of the heat of the exhaust gases to operate the choke. The engine coolant circulates through hoses to the choke water housing, where its heat is transmitted to the bimetal thermostatic spring, thus influencing the choke valve position.

Notes

Figure No. 126

ELECTRIC CHOKE-ASSIST SYSTEMS

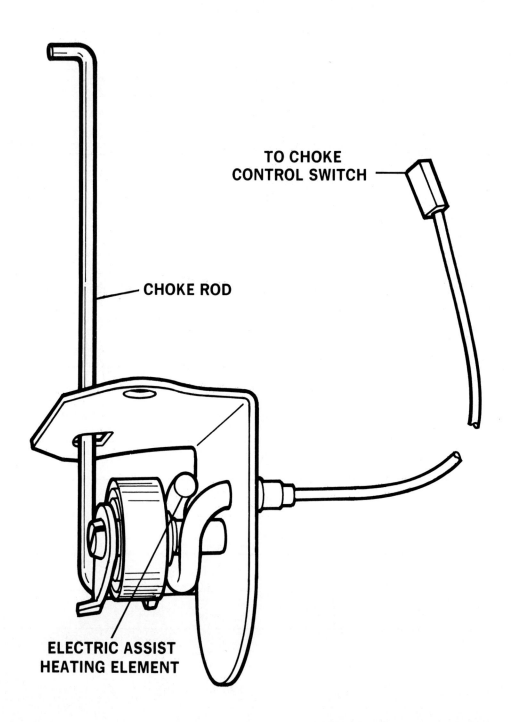

ELECTRIC CHOKE-ASSIST SYSTEMS

To hasten the warm-up process and shorten the period during which the engine is on fast idle, modern engines use an electrically heated choke mechanism. This is an electrical heating element contained within the thermostatic choke housing to supply additional heat to the thermostatic choke spring.

Most of these systems have a thermally operated switch that cuts off the heater current when the ambient temperature is below a certain level, for instance 60°F. This allows normal choke action and a longer fast idle period during cold-weather operation. In warmer weather, the cutoff switch closes to allow the heating coil to shorten the choke operating period.

Troubleshooting these systems is usually done with an ohmmeter and voltmeter. The voltmeter, by connecting it to the input terminal on the choke housing and ground, shows the presence or absence of operating voltage. In some cars, this voltage is derived from the alternator, and therefore the engine must be running for this check. Check the manufacturer's test procedure to see if this voltage is controlled by ambient temperature.

The heating element itself is generally checked by measuring its resistance with a low-range ohmmeter. The specified resistance varies, but typically is in the range of 4 to 12 ohms. Again, it is necessary to check the specified operating temperature of the choke cutoff switch. If the ambient temperature is below the operating point, the switch may be open, preventing a resistance measurement of the heating coil.

Figure No. 127

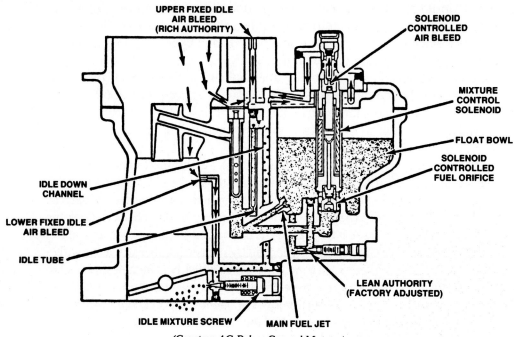

(Courtesy AC-Delco, General Motors.)

Figure No. 128

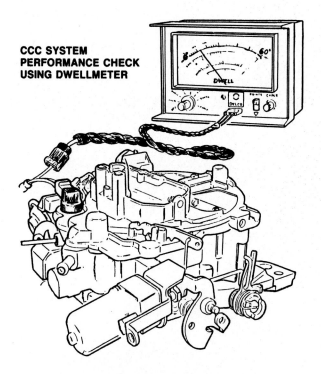

Dwellmeter Connection.
(Courtesy AC-Delco, General Motors.)

ELECTRONIC CARBURETORS

The transition to computerized engine control systems with oxygen sensor input made it necessary to apply electronic technology to the carburetor. While the basic circuits of the carburetor remain unchanged, modification of the metering circuit is needed to give the computer control of fuel mixture. As is the case with most other computer controls, a solenoid is used to effect fuel flow and air bleeding to the carburetor main-well. The computer acts as a ground base by applying or removing the ground lead of the solenoid field winding. The air-fuel mixture is controlled by the length of time that the solenoid is energized (grounded) as opposed to the length of time it is de-energized (ground removed). The computer cycles the solenoid 10 times per second.

This solenoid is referred to as a *mixture-control* solenoid or as the *feedback* solenoid. Computer commands to the mixture-control solenoid are separated into two modes: open-loop or closed-loop operation. In open-loop operation, the computer ignores the oxygen sensor, such as during wide-open throttle or cold engine conditions. While in open-loop mode, the computer maintains a constant command signal represented by a 50 percent duty cycle. A 50 percent duty cycle means that the solenoid is energized for the same length of time it is de-energized. Dwell meters are used to measure duty cycles (degrees of dwell). A dwell meter set on the six-cylinder scale has a range of 0 to 60 degrees of dwell. A 50 percent duty cycle would give a reading of 30 degrees. On General Motors vehicles, a dwell meter is used to diagnose system problems, and a constant mixture-control solenoid dwell of 30 degrees indicates open-loop operation.

During closed-loop operation, the computer uses the oxygen sensor voltage inputs to determine the mixture-control solenoid output commands. Closed-loop commands constantly vary due to the swing from rich to lean signals generated by the oxygen sensor. A lean oxygen sensor signal (low voltage) causes the computer to issue a rich command. A rich command opens the mixture-control solenoid for a longer period of time than it is held closed. The greater open duration increases fuel flow and decreases the amount of air bled into the discharge circuit. The resulting richness causes the oxygen sensor to send a rich exhaust signal (high voltage) to the computer. The computer responds by issuing a lean command to the solenoid. Lean commands mean that the solenoid is held closed longer than it is held open. Fuel flow is reduced and air bleed volume is increased. This oscillation between rich and lean commands results in a constantly varying duty cycle. A dwell meter (six-cylinder scale) should constantly swing between 10 and 50 degrees.

Figure No. 129

CARBURETOR

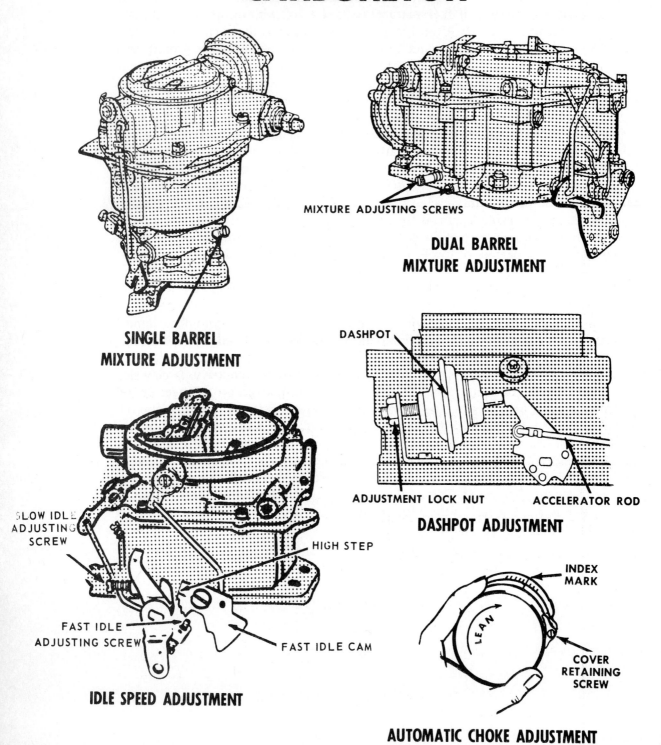

CARBURETOR ADJUSTMENT FOR NONEXHAUST EMISSION CONTROLLED ENGINES

The adjustment of the carburetor is the last step in the tune-up procedure. It is performed after all other tests and adjustments that influence engine operation have been completed. The carburetor is adjusted only with the engine at normal operating temperature.

1. Connect a tachometer and/or a vacuum gauge to the engine.
2. Start and idle the engine until normal operating temperature is reached. Be sure the engine is running on slow idle.
3. Turn the idle mixture adjusting screw in slowly until tachometer or vacuum gauge needle drops slightly.
4. Turn idle mixture adjusting screw out until tachometer or vacuum gauge returns to highest reading.
5. Repeat procedure on the other idle adjusting screw (if the carburetor is so equipped). Go back to the first adjusting screw and "trim off" the adjustment.
6. Adjust the throttle stop screw so that the engine idles at the specified rpm. Turn on the headlights or the air conditioning unit when making this adjustment, if so instructed.
7. Set fast idle adjustment to specifications.

After the carburetor is adjusted, operate the throttle linkage to check for binding and for accelerating pump discharge.

If the carburetor is equipped with a dashpot for slow throttle closing, it may be checked by allowing the throttle to snap shut several times while observing the tachometer after each check. The engine should return to the same idle speed each time. If it does not, the linkage may be sticking or the dashpot may be malfunctioning. Relieve the binding in the linkage and replace the dashpot if it does not respond to adjustment.

Hand operate the choke linkage for free operation. If sticking or binding exists, it may be caused by gum or varnish on the choke shaft or by carbon on the choke piston. The condition may be corrected by removing the carburetor air cleaner and the choke cover, and flushing the choke shaft bearings and the choke piston with solvent while hand operating the choke linkage. If the condition is not relieved, the carburetor should be removed from the engine for complete cleaning. When reinstalling the choke cover, be sure it is positioned properly in reference to the index adjusting mark (for example, 1 rich; 2 lean).

Carburetors that do not respond to adjustment must be removed from the engine for disassembly and complete cleaning.

Notes

9

EMISSION CONTROL SYSTEMS

Notes

INTRODUCTION TO VEHICLE EMISSION CONTROL SYSTEMS

The automobile engine is considered one of the major contributors to the pollution of the atmosphere. It has been established that the partially burned hydrocarbons (HC) carbon monoxide (CO), and nitrogen oxides (NO_X) contained in the automotive exhaust, when in the presence of sunlight, creates "photochemical smog." When the smog cencentrations is sufficiently high, it has definite eye and lung irritating effects. Smog is a coined word, taken from the words smoke and fog.

To combat this serious condition, federal legislation has been passed that requires every passenger car and light truck engine, beginning with the 1968 models to be equipped with federally approved emission control systems. These systems have been used, by virtue of state law, in California since 1966.

There are basically three sources of emissions from the automobile engine:

1. *Crankcase vapors:* These vapors are composed of certain amounts of compressed fuel charge and exhaust fumes that pass the piston rings as "blow-by" and collect in the crankcase. These vapors represent approximately 20 percent of the engine's emissions.

2. *Exhaust emissions:* These emissions contain (1) hydrocarbons, which are basically unburned fuel; (2) carbon monoxide, which is an invisible, odorless poisonous gas caused by overly rich fuel mixtures, for example, during periods of choke operation; (3) oxides of nitrogen. These compounds are a product of the oxygen and nitrogen in the atmosphere and are formed under the high temperatures and pressures in the engine's combustion chambers. These three by-products make up about 60 percent of the engine's emissions.

3. *Fuel evaporative emissions:* Evaporation of the fuel from the carburetor and fuel tank takes place constantly and contributes about 20 percent of the vehicle's emissions. The release of these fumes is highest during periods of "heat soak" immediately after engine shutdown. Devices and systems to control this condition were introduced on the 1970 vehicles sold in California and have been installed nationally on all cars starting with the 1971 models (since the enactment of federal control regulations).

Crankcase Emission Control

The emissions of crankcase vapors have been effectively controlled by the closed PCV system. This system is standard on all engines, nationwide, including imported cars, beginning with the 1968 models, and has been standard equipment on all California engines since 1964.

Exhaust Emission Control

The emissions from the engine's exhaust system must, by federal law, be held within specified limits. These national standards apply to all vehicles sold in the United States beginning with the 1968 models. The standards apply equally to imported car models sold in the United States.

© Copyright 1986, Tune-Up Manufacturers Institute

The federal standards established in 1968 set limits on hydrocarbon and carbon monoxide emissions. These standards have been getting progressively tighter, and is certain they will become even more stringent by 1980, at which time revised standards will have been established.

The federal government has approved several different car manufacturers' systems designed to control exhaust emissions. For purpose of identification, these systems may be basically classified into two categories: engine modification and air injection.

The engine-modification systems employ a degree of engine redesigning that permits a greater reduction of objectionable emissions within the engine's cylinders by promoting more complete combustion. In addition, these systems also employ various assist units to still further effect emission reductions.

The air-injection systems employ a belt-driving, positive displacement, low-pressure air pump that continuously pumps air through a system of hoses and tubes into the cylinder head area of each engine exhaust valve. When the exhaust valve opens, the injected fresh air supplies the additional oxygen to support further combustion of the unburned hydrocarbons in the exhausting gases. In this manner the hydrocarbons are consumed in the engine instead of being exhausted into the atmosphere. The air-injection system also incorporates various assist units to aid in controlling objectionable emissions.

Since the automobile's air-polluting emissions are highest during periods of idling and closed-throttle deceleration, most emission control devices are designed to effect these two operational areas.

Fuel Evaporation Control Systems

In addition to systems designed to control crankcase emissions and exhaust emissions, the state of California established standards to control air-polluting gasoline evaporation emission that occur from the automobile fuel tank and carburetor. This standard applied to all 1970 cars sold in California and has been applied federally to all cars beginning with the 1971 models.

Two types of evaporation control systems are used. They are essentially similar, the basic difference being the method used to store the fuel vapors. Some models use the popular system of storing the vapors in a charcoal-granule-filled canister which is mounted in or near the engine compartment. The other type of system stores the vapors in the engine crankcase. With either system, as soon as the engine is started the vapors are drawn from the canister, or crankcase, into the engine cylinders where they are burned.

The only path the fuel system vapors can follow is through the evaporation emission control system. Air pollution from this source is thereby eliminated.

Following are detailed descriptions of each of the systems discussed in this introduction.

Notes

EXHAUST EMISSION CONTROL SYSTEMS

ENGINE MODIFICATION TYPE

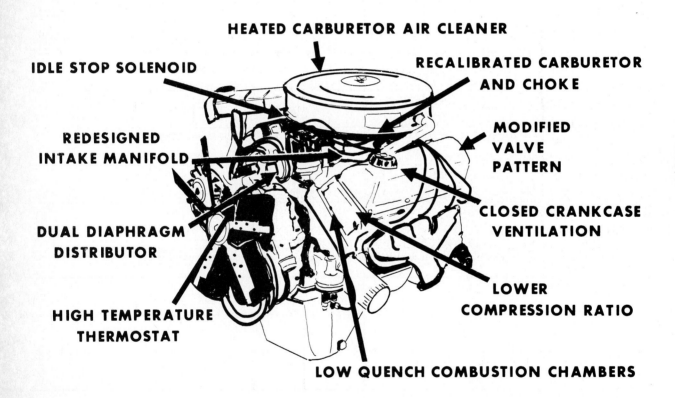

Figure No. 130

EXHAUST EMISSION CONTROL SYSTEMS

Engine Modification Type

Several exhaust emission systems come under the classification of engine modification. Each car manufacturer employing a system of this type applies his own trade name. General Motors calls its system the Controlled Combustion System (CCS); Ford Motor Company uses the Improved Combustion (IMCO) system; and American Motors calls its system the Engine-Mod (Engine Modification) System. The Chrysler Corporation Cleaner Air System (CAS), also an engine-modification type, is covered separately.

Engines employing the modification-type emission control system have been redesigned in several areas to more completely burn the fuel charge in the combustion chambers, thereby reducing objectionable exhaust emissions that are a product of incomplete combustion. Basically, this engine redesigning includes the following features. The carburetor is specially calibrated for lean fuel mixtures. The choke is more sensitive to variations in engine temperature during engine warm-up, permitting faster choke release. Higher idle speeds reduce idle emissions. Redesigned intake manifold permits smoother fuel charge movement with less restricted fuel induction, while further providing better fuel distribution.

The distributor has been recalibrated for greater spark timing control. Dual-diaphragm vacuum units provide spark advance in the cruising ranges and spark retard during idle and periods of deceleration. Ignition advance control systems (covered under assist units) are employed to prevent vacuum spark advance during acceleration and to permit vacuum advance only at cruising speeds in high gear to more effectively control emissions.

A redesigned camshaft provides a modified valve timing pattern with a greater degree of overlap in most instances. The combustion chamber contour has been modified to reduce the quench area, and the compression ratios have been reduced. In most engines the compression has been reduced sufficiently to permit the use of nonleaded gasoline. Higher temperature thermostats in the cooling system are also being used.

In addition to the engine design features listed above, these engines are also fitted with various assist units to still further control the engine's emissions.

The net result of all these improvements is more complete combustion with cleaner running engines.

Figure No. 131 **AIR INJECTION REACTOR SYSTEM (AIR)**

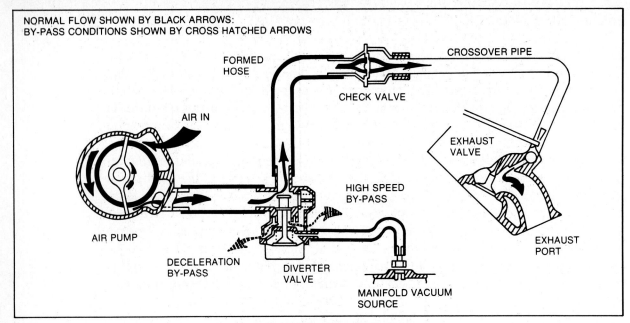

(Courtesy AC-Delco, General Motors.)

Figure No. 132 **AIR MANAGEMENT — TWO VALVE**

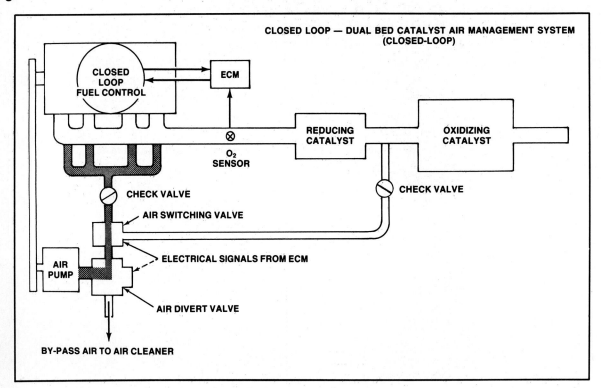

(Courtesy AC-Delco, General Motors.)

AIR INJECTION SYSTEMS

The internal combustion (IC) engine is not capable of 100 percent complete combustion, allowing small amounts of carbon monoxide (CO) and hydrocarbons (HC) to enter the exhaust. There is adequate heat in the exhaust to burn these pollutants, but an insufficient supply of oxygen exists. The air-injection system supplies the additional oxygen needed to further the burning of CO and HC in the exhaust manifolding. Air injection systems are capable of greatly reducing CO and HC output, therefore permitting the use of richer mixtures and greater ignition advance for improved drivability and performance.

Air-Pump-Type Systems

Air-pump-type systems, (Figure No. 131) are composed of four major components: air pump, diverter valve, air distribution manifold, and check valve.

The *air pump* is a belt-driven vane-type compressor. It is mounted to the engine with an accessory drive bracket similar to that of an alternator or A/C compressor. Adjustment of drive-belt tension is the same as for an alternator.

There are three methods of filtering inlet air to the pump. Imports favor the use of a small pleated paper filter attached to the air-pump intake port. Some domestic manufacturers draw air from the filtered air section of the carburetor air cleaner assembly. By far the most popular filtration method is the centrifugal filter/fan assembly. The intake fan is fitted to the pump drive shaft behind the pump pulley. Since dirt particles are heavier than air, the spinning fan centrifugally separates air and dirt. The filtered air enters the pump through an intake port located behind the filter/fan.

Engine speed determines the volume of air injected. Injection air pressure is limited by a pressure-relief valve. The pressure-relief valve can be located on the air pump housing or in the diverter valve. When pressure exceeds the predetermined value, such as during high engine speeds, the valve opens, releasing the excess pressure into the atmosphere. The maximum air injection pressure is generally limited to less than 5 psi.

Aside from belt tension adjustment and intake filter replacement (if so equipped), no routine maintenance is required. The most common failure is a seized pump impeller, which causes the belt to snap. Since the pump cannot be rebuilt, it must be replaced.

The *diverter valve* is responsible for directing pump air to either the exhaust or to atmosphere. If air were injected into the exhaust during deceleration, severe backfire would occur with possible damage to the exhaust system. The introduction of air into the superrich deceleration exhaust stream creates an expansion rate greater than the exhaust plumbing can handle. Therefore, the diverter valve must direct pump air to the atmosphere.

Diverter valve operation is controlled by manifold vacuum. Deceleration generates very high manifold vacuum levels (greater than 20 inches of mercury). This high vacuum is applied to the diverter valve diaphragm, which overcomes a calibrated spring tension moving the valve plates (or piston) into an atmospheric dump position. When manifold vacuum drops below a predetermined level, the diaphragm assumes a normal operation position and air is diverted to the exhaust header.

© Copyright 1986, Tune-Up Manufacturers Institute

Understanding diverter valve operation makes it easy to test. Simply raise the engine speed to 2500 to 3000 rpm and let the throttle snap closed. Air should be felt (or heard) escaping from the valves' atmospheric vents for approximately 2 to 3 seconds. The valve cannot be serviced and must be replaced if it fails testing.

The *check valve* is located in the air distribution manifold between the diverter valve and injection nozzles. The check valve allows air to pass through to the nozzles, but prevents exhaust gases from reaching the diverter and pump. In the event of check valve failure, heat and exhaust deposits would rapidly destroy the diverter valve and air pump.

To test the check valve, remove the air-supply hose from the valve. While the engine is idling, no exhaust gases should escape from the inlet side of the valve. The sound made by the open valve will resemble a muffler leak and is normal. Remember, some V engines have two check valves, one for each bank of cylinders.

Aspirator Systems

Aspirator-type systems are commonly called *pulse air injection*. These systems are generally used on smaller engines with initially low HC and CO outputs. No air pump is used because the pulse air injection operating principle relies on negative pressure pulses in the exhaust system. Due to the high velocity of exhaust charges in the exhaust, pressure pulses from positive to negative several times per second. During the negative pulses, a reed-type check valve opens, allowing filtered air to enter the header. When positive pulses occur, the check valve closes and prevents exhaust heat and gases from damaging the air cleaner. The small amount of air delivered by these systems is adequate for reducing marginal HC and CO levels and is helpful in supporting the oxidation process of a catalytic converter.

Air-Management Systems

Vehicles equipped with computerized engine controls and dual-stage catalytic converters use a complex air-management system (Figure No. 132). While the air pump and check valves remain the same as conventional air-pump systems, the diverter valve is replaced with an electronic air-management valve. The air-management valve controls the flow of air in three directions: to atmosphere, upstream of the oxygen sensor, or downstream of the oxygen sensor to the catalytic converter.

The computer energizes or de-energizes a pair of solenoids, which in turn controls vacuum signals to different sections of the air-management valve. The computer receives information from various engine sensors: coolant temperature, throttle position, and manifold vacuum.

The air-control section of the valve directs air-pump discharge to either the air switching valve or to atmosphere. The computer turns the air-control solenoid on and off, regulating vacuum signals to the air-control diaphragm. Ford calls this solenoid the thermactor air by-pass solenoid (TAB). The function of the air-control section is very similar to the old diverter valve.

The air switching section sends air from the air-control valve to either upstream or downstream of the oxygen sensor. Again, the computer grounds or ungrounds a solenoid to control the vacuum signal to the air switching diaphragm. Ford denotes this solenoid as the thermactor air divert (TAD) solenoid. Under normal operating conditions, the air switching valve should send air only to between the converter beds (downstream). During cold engine operation, the air switching valve will allow air to enter the exhaust header. This upstream air assists in heating of the oxygen sensor and light-off of the catalyst, in order to put the computer into closed-loop operation as soon as possible.

A malfunction in the air switching valve can have severe effects on the operation of the loop (oxygen sensor, computer, mixture-control solenoid). If the valve were to inject air into the header (upstream) during closed-loop operation, the oxygen sensor would detect high oxygen content in the exhaust and signal the computer that a superlean condition exists. Thus the computer would issue a rich command to the mixture-control solenoid. The combination of rich mixture and ample oxygen supply causes rapid catalytic converter meltdown and subsequent exhaust restriction. The causes of this malfunction range from simple valve leakage to complex failures in the sensory and/or computer systems.

Figure No. 133

HEATED CARBURETOR AIR SYSTEMS

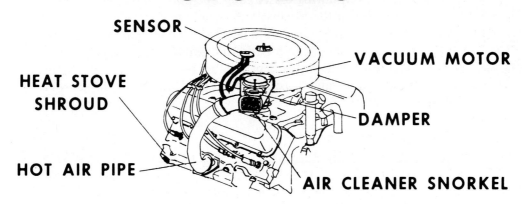

GENERAL MOTORS AND CHRYSLER TYPE

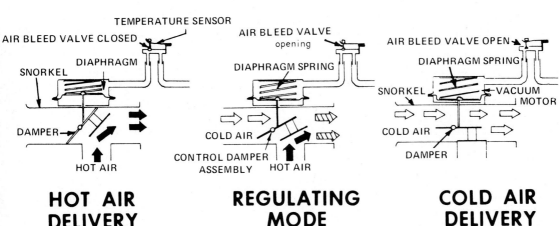

HOT AIR DELIVERY — REGULATING MODE — COLD AIR DELIVERY

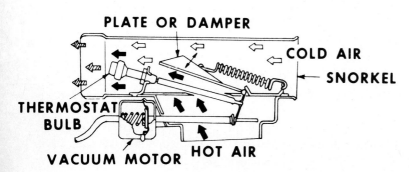
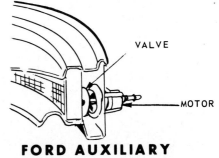

FORD AND AMERICAN MOTORS TYPE

© Copyright 1986, Tune-Up Manufacturers Institute

HEATED CARBURETOR AIR SYSTEMS

The heated carburetor air system (also called "Hot and Cold" Air Intake system or Climatic Combustion Control) has been introduced as another means of assisting in the control of objectionable exhaust emissions.

The function of the system is to direct heated air to the carburetor when underhood temperature is below 100°F and up until the temperature reaches approximately 130°F. This system provides desirable emission control throughout the vehicle's operating range, while further effecting increased fuel economy and improved engine warm-up.

Heating the carburetor air during periods of cold-engine operation permits proper fuel vaporization with more complete combustion of the fuel mixture. It further provides for a shorter period of choke action, which also has the benefit of improving engine warm-up and lowering exhaust emissions. Carburetor icing, a troublesome stalling condition prompted by the buildup of ice crystals on the throttle plate, is virtually eliminated by the intake of heated air.

A temperature-sensing device in the carburetor air cleaner housing, or snorkel, senses the incoming air temperature and through suitable linkage operates a valve plate or damper door. When underhood temperature is below 100°F, the valve plate is held closed (heat-on position) so that only heated air, drawn through a duct connected to a heat stove positioned around the exhaust manifold, can enter the carburetor. As engine and underhood temperatures rise, the sensing device opens the valve plate slightly, permitting a mixture of heated and unheated air (at about 100°F) to enter the carburetor. As operating temperature rises to normal (130°F or higher), the sensing device opens the valve plate completely (heat-off position) so that only engine compartment unheated air enters the induction system.

The temperature-sensing device can be a *vacuum-control* unit which governs the vacuum motor action of controlling the damper door by applying or bleeding off vacuum to the motor. General Motors and Chrysler vehicles employ this type. Or the device can be a thermostat capsule or bulb which *mechanically* operates the damper door. This is the type used on Ford and American Motors engines.

To insure acceptable engine performance during periods of cold-engine acceleration when the supply of warm air may not be adequate, the damper must be momentarily fully opened to unheated air to provide an air supply that is sufficient to the engine's demands. In the General Motors and Chrysler vacuum-operated unit, the drop in intake manifold vacuum on sudden acceleration permits the vacuum motor diaphragm spring to open the damper. On Ford and American Motors mechanically operated units, an auxiliary air inlet valve and the vacuum motor that operates it are mounted on the rear of the air cleaner housing. The sudden drop in manifold vacuum that accompanies quick throttle opening permits the valve to open and provide the additional air required for acceptable acceleration. Earlier Ford models employed a vacuum override motor (mounted beneath the air cleaner snorkel) that momentarily opened the valve plate by the action of a piston rod.

Air entering the carburetor, either from the heat stove or from the underhood area, passes through the air cleaner element where the dust and dirt particles contained in the air are removed.

On dual-snorkel heated air cleaners, one snorkel is connected to the temperature sensor and draws air through the heat stove exactly as does the single snorkel air cleaner previously described. The other snorkel functions strictly by manifold vacuum.

© Copyright 1986, Tune-Up Manufacturers Institute

The two vacuum motors are connected by a T in the vacuum hose connected to the carburetor. During periods of hard acceleration, manifold vacuum drop permits both vacuum motors to open the snorkels to underhood air, permitting sufficient air intake for maximum performance. During periods of high-speed operation, the increased air velocity causes a pressure drop inside the air cleaner. With a higher pressure on the outside of the door, the pressure differential tends to open both snorkel doors to underhood air regardless of manifold vacuum.

On many late model cars, the application of heated air to the carburetor on cold engine starts has also led to the elimination of the exhaust manifold heat control valve sticking problems.

Notes

Figure No. 134

HEATED CARBURETOR AIR SYSTEM TESTS

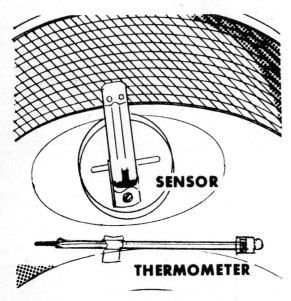

TESTING TEMPERATURE SENSOR

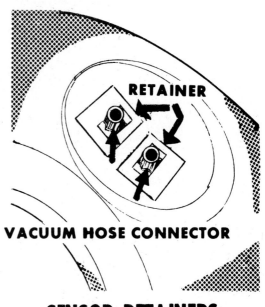

SENSOR RETAINERS

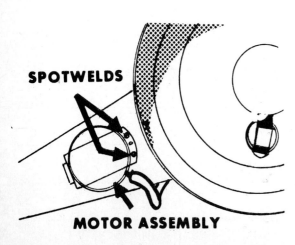

VACUUM MOTOR SPOTWELDS
GM type

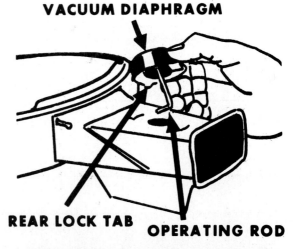

VACUUM MOTOR REMOVAL
Chrysler type

© Copyright 1986, Tune-Up Manufacturers Institute

TESTING HEATED CARBURETOR AIR SYSTEMS

There are certain recognizable symptoms that arise when the carburetor heated air system malfunctions. A cold engine will idle roughly, and performance will be erratic due to the fact that cold air, rather than warm air, is entering the carburetor. On the other hand, poor high-speed performance and power loss will be experienced if warm air is applied to the induction system after the engine has reached operating temperature.

General Motors, Chrysler, and Ford Units

A quick check to determine if the automatic air temperature device is functioning properly is to look into the air cleaner snorkel before the engine is started. The damper door should be open so the air cleaner element should be visible. A mirror may be required depending on the air cleaner installation. If the snorkel passage is closed, trouble in the damper is indicated.

The vacuum-bleed sensor unit can also be easily tested. Remove the air cleaner cover and tape a small thermometer next to the sensor unit. The tape will prevent the thermometer from being sucked into the engine when it is started. Lay the cover back on the cleaner. With a cold engine at idle speed, the damper door should be closed. (if the temperature is above 85°F, the damper door will be partially open, and it will only be necessary to make certain that the damper is completely opened at 130°F). Observe the action of the damper in the snorkel as the engine warms up. When the damper begins to open, lift the air cleaner cover and check the thermometer reading. At a temperature of about 85°F the damper should start to open, and at approximately 130°F it should be fully opened.

Since the temperature sensor is preset and is nonadjustable, a malfunctioning unit must be replaced. It is held in place by two flat retaining clips. The new unit and gasket must be placed in the air cleaner housing in the same relative position as the original unit.

To check the operation of the vacuum motor, connect a test vacuum hose from the intake manifold to the motor. With the engine idling, the damper should close the cold air passage from the snorkel. A defective motor can be removed by drilling out the retaining spot welds on the General Motors air cleaner and securing the new motor in place with self-tapping sheet-metal screws. The Chrysler vacuum diaphragm is removed by bending down the front lock tab, lifting the front edge of the motor, disengaging the rear tab, and then unhooking the operating rod from the heat control door.

Ford and American Motors Units

As an initial check on a cold engine with underhood ambient temperature of less than 100°F, look into the air cleaner snorkel. The valve plate should be in the heat-on or up position. If the valve is not in the proper position, remove the duct and valve assembly and submerge it in a container of cool water. With a thermometer placed in the water, slowly heat the water. At a water temperature of 100°F or colder, the valve should be in the heat-on or up position. If the valve does not move, check for possible mechanical interference of the plate in the duct.

© Copyright 1986, Tune-Up Manufacturers Institute

Raise the water temperature to approximately 110°F, allow a few minutes for the temperature to stabilize, and the valve should start to open. Increase the water temperature to approximately 135°F and again allow for temperature stabilization. The valve should now be in the heat-off or down position. If the valve does not conform to these requirements, replace the unit with a new valve and duct assembly.

The override vacuum motor can be checked in the following manner. With a cold engine and underhood temperature at less than 100°F, start the engine and observe the position of the valve plate. It should be in the heat-on or up position. If it is not, remove the hose from the vacuum motor and check the available vacuum at the hose. It should be at least 15 inches at idle speed. If the vacuum is less than 15 inches, check for vacuum leaks in the hose and hose connections. If the proper amount of vacuum is applied to the motor, but the valve plate is not in the heat-on position, the motor is defective and should be replaced.

The Ford auxiliary air inlet valve should be fully closed with the engine idling. Disconnecting the motor vacuum hose should permit the valve to open fully. To check for faulty operation, check for interference caused by misalignment of the valve plate or motor rod. Then test the vacuum applied to the motor. There should be a minimum of 15 inches. If the vacuum is below this figure, check for hose and connection leaks. If vacuum is sufficient but the valve remains in one position, remove the vacuum motor from the air cleaner housing and test its operation at any other vacuum source. Replace the motor if it is defective.

Service Tip

When removing the heated carburetor air cleaner from the engine, it is important to remember that a vacuum hose is connected to the underside of the cleaner housing. The hose from the closed PCV system may also be attached to the air cleaner housing. It is essential, of course, that these hoses be reconnected when the air cleaner is replaced on the carburetor.

Be careful not to damage the plastic fittings which are frequently used in connection with vacuum hoses.

Notes

Figure No. 135

IDLE STOP SOLENOIDS

GENERAL MOTORS AND CHRYSLER TYPE

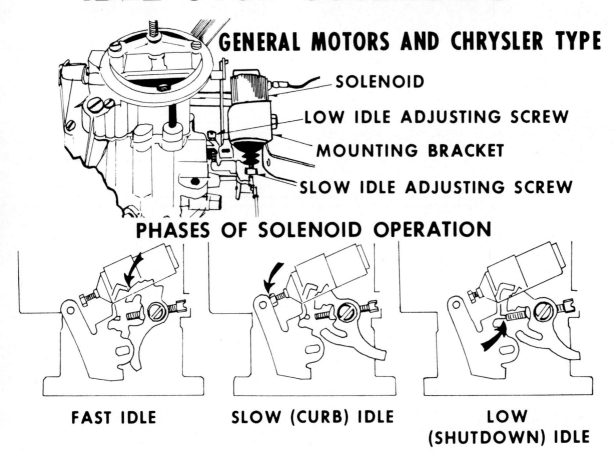

FORD SOLENOID THROTTLE MODULATOR

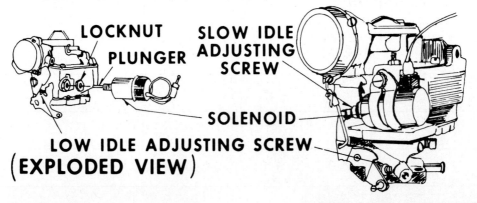

IDLE STOP SOLENOIDS

The idle stop (or antidieseling) solenoid is used on the engines of many late model vehicles. Most exhaust emission controlled engines idle at higher speeds, use leaner fuel mixtures, with retarded ignition timing, and have higher than average cooling system temperatures than engines that are not emission controlled. Further, the idle speed of the engine is set with the transmission selector in *drive* position. When the driver places the selector in *neutral,* before shutting the engine *off,* the idle rpm increases when the transmission load is removed. All these factors increase the tendency of the engine to "run on" or "diesel" after the ignition switch has been turned *off.* Because the engine keeps running on only a few of its cylinders, run-on is usually very rough and is frequently associated with violent detonation. The function of the idle stop solenoid is to prevent this tendency by permitting the idle speed to drop low enough to virtually shut off the engine's air supply, thus causing the engine to stop running immediately when the ignition is turned *off.*

General Motors and Chrysler Solenoid

The idle stop solenoid is bracket mounted on the carburetor and works in the following manner. When the ignition switch is turned *on,* the solenoid plunger is electrically activated and moves to an extended position. This action places the adjusting screw the plunger carries in contact with the throttle linkage, where it serves as the idle speed control. When the ignition switch is turned *off,* the plunger is deactivated and retracts into its housing. Since the plunger adjusting screw now ceases to serve as the throttle stop, the throttle linkage moves to a lower idle setting established by the carburetor idle speed adjusting screw.

On the Chrysler solenoid the adjusting screw is mounted on the linkage contacting the solenoid plunger, rather than on the plunger itself. The idle adjusting screw and solenoid further serve as a switch to activate the distributor solenoid, on engines so equipped. The function of the distributor solenoid is to provide spark retard during idle and closed throttle deceleration periods. This subject is covered in the distributor section of this manual.

Figure No. 135 illustrates the three phases of solenoid operation. When a cold engine is started, it runs at a fast idle on partial choke. The solenoid is activated and the plunger is extended, but it cannot reach the throttle linkage because the fast idle cam is keeping the engine running at a speed higher than the one for which the solenoid is set. At normal operating temperature (choke open), the solenoid plunger governs the engine idle speed by acting as the throttle stop. The plunger remains extended during all periods of engine operation. With ignition switch turned *off* and the engine shutting down, the solenoid is deactivated and the plunger draws back into the solenoid housing. The engine rpm is now controlled by the carburetor slow idle adjusting screw. This speed is low enough to prevent run-on, but just high enough to prevent the throttle plate(s) from closing completely and scuffing the throttle bore(s).

On warm engine start-up, there will be, of course, no fast idle cam action. The solenoid plunger will immediately serve as the throttle stop. But here is an important point. The solenoid, though activated when the ignition switch is turned *on,* may not be strong enough to open the closed throttle against the weight of the linkage and the action of the return spring. After turning on the switch, it is necessary to depress the accelerator approximately one-third of its travel to permit the plunger to move to its extended position, where it will then be able to hold the throttle open. Failure of the engine to continue running after start-up may be experienced if this procedure is not observed.

When starting a cold engine, the accelerator must be fully depressed to set the choke and fast idle cam as well as the solenoid.

Ford Motors Throttle Modulator

When used on Ford-built six-cylinder engines, the idle stop device is called a *solenoid throttle modulator*. It serves the same function and operates in the same manner as previously described.

The throttle modulator used on the Carter YF carburetor is adjusted for setting the slow (curb) idle by turning the adjusting screw as mentioned above. The solenoid used on the Autolite carburetor is adjusted by loosening the locknut and turning the solenoid in or out to obtain the specified slow idle and then tightening the locknut. The Solenoid on some late-model engines is mounted on an adjustable bracket slide, providing a means of setting the solenoid adjustment with the air cleaner installed.

When setting the low (shutdown) idle on Ford solenoid throttle modulators, the solenoid is electrically disconnected at the bullet connector in the harness lead, not at the solenoid, as is the General Motors unit.

Three Idle Speeds

With the introduction of the idle stop solenoid (or throttle modulator), there also came the third idle speed setting. Now there is a fast idle, a slow (curb) idle, and a low (shutdown) idle.

The cold fast idle is set by a throttle linkage fast idle adjusting screw (or by bending a fast idle cam follower) to a specified speed which will be high enough to keep a cold, partially choked engine running without stalling or loading.

The slow (curb) idle is set by adjusting the solenoid plunger adjusting screw (GM and Ford), by setting the adjusting screw contacting the solenoid plunger (Chrysler), or by turning the solenoid or adjusting the solenoid bracket (Ford) to the recommended idle speed setting.

The low (shutdown) idle is set by adjusting the carburetor low idle adjusting screw after electrically disconnecting the solenoid lead at the solenoid (GM) or at the bullet connector at the loom (Ford). The Chrysler low idle is set with the solenoid energized and the engine idling by adjusting the carburetor idle speed screw inward until the tip of the screw just touches the top on the carburetor. Then back the screw out one full turn to obtain a slow idle of approximately the desired setting.

The three idle speeds may be, for example: a fast idle of 2000 rpm, a slow idle of 750 rpm, and a low idle of 500 rpm. Setting all three idle speeds to specifications is an important part of the tune-up procedure.

Notes

Figure No. 136

HOT IDLE COMPENSATOR VALVE

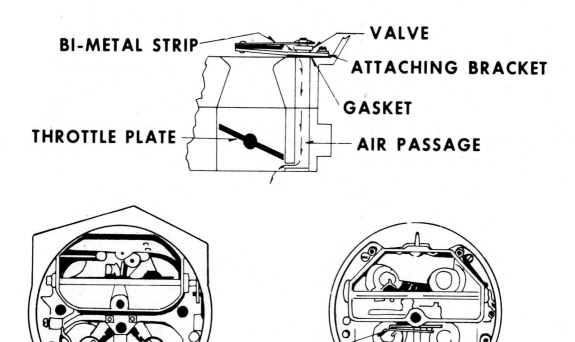

INTERNAL CARBURETOR MOUNTINGS

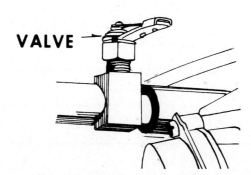

EXTERNAL MOUNTING IN PCV SYSTEM HOSE

HOT IDLE COMPENSATOR VALVE

During periods of high underhood temperatures, as in start-and-stop driving in congested traffic on a hot day, there is a tendency for fuel vapors to collect in the intake manifold. These vapors are the result of fuel "spillover" caused by the pressure generated by the heat on the fuel in the carburetor bowl. The result is an overly rich fuel mixture, which prompts rough idle and a tendency to stall. Engines that stall under this condition are frequently very difficult to restart. The function of the carburetor hot idle compensator valve is to prevent engine stalling under these conditions by opening an auxillary air bleed circuit. This action permits additional air to enter the manifold under the throttle plates and dilute the rich fuel mixture in the manifold.

The hot idle compensator valve is a simple mechanism mounted internally in the carburetor air horn on most four-barrel and many two-barrel carburetors. Some two-barrel carburetors have the hot idle valve externally mounted in the PCV hose near the carburetor. The valve, positioned over its port, is mounted on and controlled by a flat thermostatic spring. During periods of normal operating temperatures, the valve remains closed and there is no action in the idle compensating. When the air temperature in the carburetor air horn reaches approximately 120°F, the thermostatic spring opens the valve, permitting additional air to bypass the throttle plates and dilute the rich fuel mixtures collecting in the intake manifold. When operating temperatures return to normal, the compensating valve automatically seats itself, and the air bleed circuit is shut off.

The compensating valve in most carburetors is internally mounted in a horizontal position in the carburetor throat. In other carburetors, Ford and Autolite particularly, the valve is mounted on its side, so to speak. In any event, the valve is quite obvious when its size, shape, and function are understood.

The hot idle compensator valve is nonadjustable, and no attempt should ever be made to bend the thermostatic spring as a method of adjustment. Valve replacement is the only recommended procedure in the event of valve malfunction.

In most carburetor adjustment procedures there are several qualifying requirements, among which is the statement "Hot idle compensator valve held closed." Since the carburetor adjustment is one of the last settings made during a tune-up and the engine may have been idling for some time, the hot idle compensator valve may be open.

If the carburetor is adjusted with the valve open, the mixture adjustment will be made to compensate for this additional air supply. When the engine temperature returns to normal and the valve closes, the idle mixture will be too rich and the idle will be rough. To prevent this condition from occurring, the valve (if it is open) should be held closed with a gentle pressure on the valve using a pencil or a fingertip, or similarly blocking off its air passage, while the carburetor is being adjusted.

Figure No. 137

DECELERATION VALVE

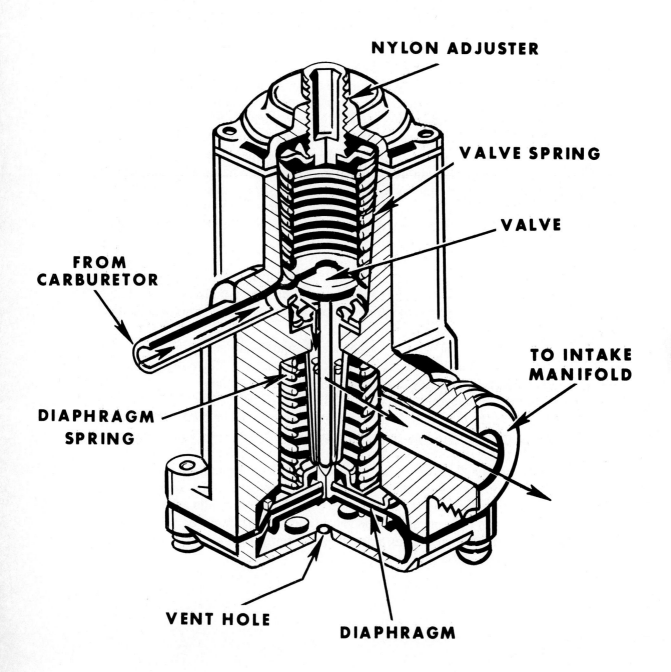

DECELERATION (DECEL) VALVE

Some Ford four-cylinder engines are fitted with an assist emission-control device called a *deceleration valve*. The valve was introduced on some 1971 models.

The generation of objectionable hydrocarbon and carbon monoxide exhaust emissions is greatest during periods of deceleration. The function of the deceleration valve is to assist in controlling these emissions by metering additional amounts of fuel and air into the engine during these periods.

The vacuum-operated deceleration valve is mounted on the intake manifold adjacent to the carburetor. The valve housing contains a spring-loaded valve in its upper chamber and a spring-loaded diaphragm in its lower chamber. The bottom cover contains a bleed hole to permit the constant entry of atmospheric pressure to the lower side of the diaphragm.

A nylon or plastic adjusting screw positioned in the top cover provides a means of setting spring tension that governs both the valve opening time and its duration of opening.

One hose connects the deceleration valve to the *deceleration section* of the carburetor. This section is cast in the carburetor's upper housing and consists of a fuel pick-up tube and an air bleed tube. Both tubes have precision-drilled restrictors and a common outlet tube to which the deceleration valve hose is connected. Another hose connects the valve to the intake manifold.

Only during periods of engine deceleration is the intake manifold vacuum strong enough to lift the deceleration valve diaphragm against spring tension, thus lifting the valve from its seat. With the valve open, intake manifold vacuum draws a metered amount of fuel and air from the carburetor deceleration section. The air-fuel mixture passes through the valve and into the intake manifold. This additional fuel and air provides a greater volume of combustible mixture in the engine cylinders, producing a more rapid and complete combustion process. Reduced levels of exhaust emissions are thereby reached.

The period of additional fuel feed time is calibrated to last for about 3 to 5 seconds, after which time the combined action of the drop in manifold vacuum and the tension of the diaphragm spring and the valve spring closes the valve.

EXHAUST GAS RECIRCULATION (EGR) SYSTEMS

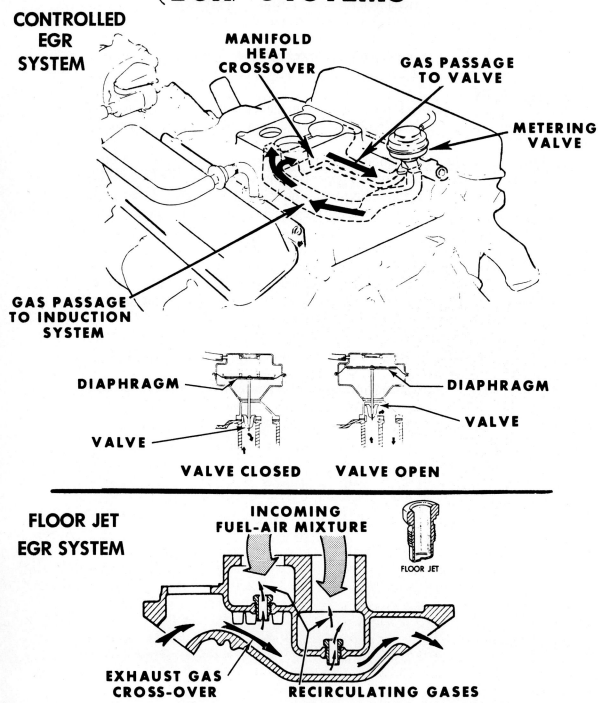

Figure No. 138

EXHAUST GAS RECIRCULATION (EGR) SYSTEMS

The exhaust gas recirculation systems introduced on some 1972 car models are designed to control the formation of the various oxides of nitrogen (NO_x) in exhaust gases.

The high temperatures and lean air-fuel ratios that are instrumental in the control of hydrocarbon (HC) and carbon monoxide (CO) emissions unfortunately are conducive to increasing the formation of nitrogen oxide emissions. It therefore became necessary to design a system that would effectively limit the formation of oxides of nitrogen without unfavorably influencing the control of HC and CO emissions. The exhaust gas recirculation system serves this important function.

The EGR system is based on the principle that the formation of oxide of nitrogen emissions can be limited by reducing the peak temperatures of the burning fuel charge in the engine's combustion chambers. This can be accomplished by the introduction of an inert material into the fresh fuel charge; since exhaust gas is an inert material and is in plentiful supply, it is used in the EGR system for induction into the intake manifold.

Diluting the fresh fuel charge with a calibrated amount of exhaust gas was formerly accomplished by the engine valve timing pattern that increased the degree of valve overlap. By having both the intake valve and exhaust valve open in the same cylinder at the same time for a limited number of degrees, the intake charge was somewhat diluted by the last of the exhausting gases leaving the cylinder. When the need for more fuel charge dilution than the valve overlap period could provide became a requirement, the exhaust gas recirculation system was devised.

Controlled Exhaust Gas Recirculation System

The controlled exhaust gas recirculation system is composed of a valve and two passages cast in the intake manifold. One passage leads from the manifold heat crossover passage to a metering valve. The other passage leads from the valve to centrally positioned holes in the intake manifold floor below the carburetor.

The metering valve is normally held closed by a coil spring positioned above the valve diaphragm. Diaphragm movement is under the influence of a vacuum signal initiated at a point above the carburetor closed throttle plates. At idle speed, the very weak carburetor vacuum cannot overcome the diaphragm spring tension, and the valve remains closed. Dilution of the fuel charge is prevented and a smoother idle is maintained. Since the formation of nitrogen oxides is lowest at idle speed, dilution of the fuel charge is not essential at idle.

As the throttle is opened, the formation of nitrogen oxides increases rapidly. To control this condition, the opening throttle plates expose the metering valve diaphragm to increased intake manifold vacuum, the valve is lifted off its seat, and a metered amount of exhaust gas is permitted to enter the induction system to limit the formation of NO_x emissions, precisely when needed.

Certain components have been added on some systems to more accurately control EGR operation. Since intake vacuum is also very low at full throttle, various devices are used to help supply the needed vacuum for more precise EGR control. The *vacuum amplifier* stores manifold vacuum in a reservoir, which is released to the EGR valve upon signal under full-throttle conditions. The *back-pressure transducer valve* adjusts

EGR flow according to exhaust back pressure. The higher back pressure which occurs during full throttle activates this system to supply more vacuum to the EGR valve.

Various devices are also used to help cold-start driveability. *Time delay valves* hold off recirculation for a short period of time after the engine has been started. *Ambient temperature switches,* either air-temperature or coolant-temperature activiated, bleed or black off the vacuum signals according to surrounding underhood temperature, usually about 63°F or below.

The controlled EGR valve can be checked to determine if the valve opens and closes properly. Procedures vary, but the general method is to operate the engine at a fast idle and then remove and reconnect the valve's vacuum line while noting the change in speed. When the vacuum line is off, the speed should increase, typically to about 50 to 100 rpm. The engine should return to its original speed when the vacuum line is reconnected. This indicates that the valve closes (speed increases) when the vacuum is interrupted, and opens (speed decreases) when normal carburetor vacuum is applied to the valve. If the test speed is too low, normal EGR or carburetor vacuum will not be present, and the test will be inconclusive. At idle speeds, no vacuum is applied to the EGR valve, thereby keeping it closed.

This EGR system is classified as *variable* since the control valve action determines when and to what extent fuel charge dilution occurs.

Floor Jet Exhaust Gas Recirculation System

The floor jet EGR system provides for the entrance of exhaust gas into the intake manifold of V-8 engines through metered orifice jets positioned in the floor of the manifold heat crossover passage below the carburetor. The orifice jet in six-cylinder engines is located in the manifold "hotspot" below the carburetor heat riser.

The calibrated size of the orifice in each jet determines the amount of exhaust gas that can be inducted into the intake manifold by engine vacuum.

The jets are made of nonmagnetic steel and are threaded into the floor of the intake manifold so that they can be removed for cleaning of the orifice, should the need occur.

Checking EGR Systems

This EGR System is classified as *fixed* since the amount of fuel dilution is controlled by the fixed size of the orifice in the jet.

The floor jet EGR system can be checked only by visual examination after removing the carburetor. If the floor jets, located below the carburetor, appear clogged or otherwise damaged, they must be replaced. They must be removed carefully since, being made of stainless steel, they are nonmagnetic and difficult to retrieve if dropped.

Notes

Figure No. 139

DUAL DIAPHRAGM EGR VALVE

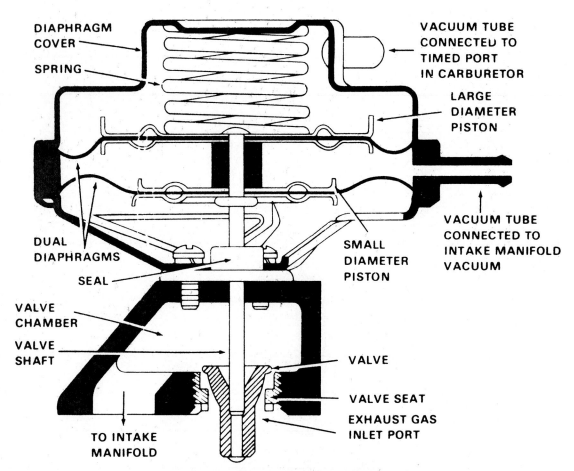

**DUAL DIAPHRAGM
EXHAUST GAS RECIRCULATION VALVE**
(PARTIALLY OPEN-CRUISE)

DUAL DIAPHRAGM EGR VALVE

A modified version of the EGR valve shown previously will also be found. This is the dual diaphragm or proportional-type valve. Its purpose is to provide more precise control over the amount of exhaust gas admitted to the incoming mixture so as to provide improved engine operation and driveability.

In principle, it is similar to the single diaphragm valve. However, the additional diaphragm makes the valve more responsive to changes in intake manifold vacuum. These valves have two vacuum lines connecting to it. One goes to the vacuum port on the carburetor, the same as on the single diaphragm valve. A second vacuum line connects to the intake manifold.

Figure No. 140

POSITIVE BACKPRESSURE EGR VALVE

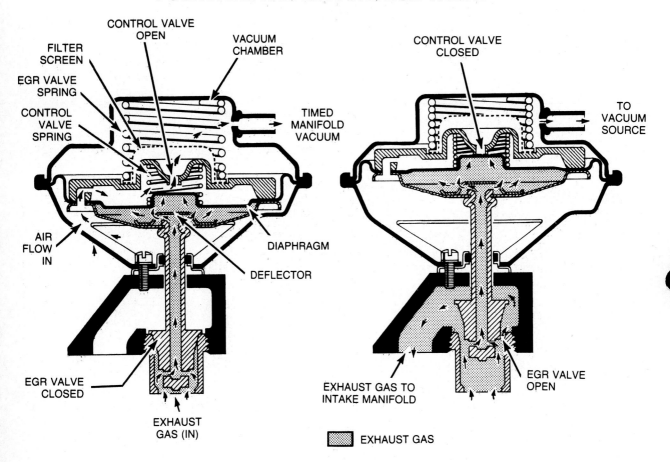

(Courtesy AC-Delco, General Motors.)

BACKPRESSURE TRANSDUCED EGR VALVES

In the latter half of the 1970s, most domestic manufacturers adopted backpressure transduced EGR systems. The earlier ported vacuum EGR valves caused drivability problems because they were not capable of metering intake charge dilution in relation to exhaust pressure. During periods of high exhaust pressure (load), excessive amounts of exhaust gases would be introduced into the intake, resulting in sudden losses of power (hesitation and flat spots). There are two types of backpressure valve construction: remote sensor and integral sensor.

Remote-Sensor-Type Backpressure EGR Valve

The remote exhaust pressure sensor is spliced into the vacuum signal line between the EGR valve and the temperature vacuum switch. The EGR valve vacuum signal is supplied by ported vacuum, just as on the conventional EGR valve systems. A steel tube exits the bottom of the backpressure sensor, connecting the sensor to exhaust pressure in the crossover section of the intake manifold. The internals of the sensor consist of a silicone rubber diaphragm, a calibrated spring, and a vacuum bleed port. These components are assembled in such a way as to make the sensor a positive exhaust backpressure transducer.

    The function of the sensor is to modify the ported vacuum signal to the EGR valve in relation to exhaust pressure. At idle and light load, exhaust pressure is very low. With the throttle plate in an idle position, there is no vacuum signal for the sensor to affect, but at light and initial load the vacuum signal is very strong. This strong vacuum signal would fully open the EGR valve on a conventional system, but not on a transduced system. Since the exhaust pressure is not high enough to force the diaphragm against the bleed port, most of the ported vacuum is vented to atmosphere. The bleeding off of ported vacuum weakens the signal, resulting in an EGR valve that opens very little or not at all.

    Under moderate load or cruise operation, the exhaust pressure is high enough to force the diaphragm upward against the bleed port, preventing the entry of atmospheric pressure into the vacuum line. Ported vacuum is strong enough to open the EGR valve, allowing exhaust gas recirculation. Because the opening of the EGR valve causes pressure changes in the exhaust pressure signal, the valve begins to modulate. The exposure of manifold vacuum to the exhaust crossover passage reduces the backpressure signal and allows the sensor diaphragm to uncover a portion of the bleed port. The bleeding-off process weakens the ported vacuum signal, and the EGR valve closes until pressure rises again. The subsequent rise in pressure closes the bleed and the EGR valve opens, beginning the modulation again. Thus a proper operating system will have the EGR valve opening and closing several times per second.

    Wide-open throttle (WOT) produces high exhaust pressure and low manifold vacuum. While the crossover pressure is more than capable of closing off the bleed port, the ported vacuum signal is not adequate to lift the EGR. Thus no EGR exists during hard acceleration, allowing maximum combustion chamber heat and power development.

    Since 1980, most manufacturers have stopped using remote-sensor-type EGR valves and have adopted the *integral transducer valve*. Instead of using two separate components, the sensor is located within the EGR valve. This type of construction is less expensive to manufacture and reduces assembly line time. The EGR pintle stem is hollow in order to pass crossover exhaust pressure to the transducer diaphragm. When exhaust pressure is high enough, the transducer diaphragm is pushed upward against the control-valve spring, covering the bleed hole in the main EGR valve diaphragm. When the bleed is closed, the vacuum signal can lift the EGR valve open. Control and operation are the same as for the divorced-sensor-type.

Negative Backpressure Transduced EGR Valves

The negative backpressure valve is simply a variation of the previously described positive backpressure types. It is an integral-type design with the only difference being the location of the control-valve spring. The spring is located beneath the transducer diaphragm instead of above it. When ported vacuum partially opens the valve, manifold vacuum is transmitted through the hollow pintle stem. This limited vacuum signal draws the transducer diaphragm downward, uncovering the air bleed and enabling the EGR valve to modulate. The modulation maintains EGR flow at a constant percentage of intake air flow.

Notes

Figure No. 141

ECM CONTROL OF EGR

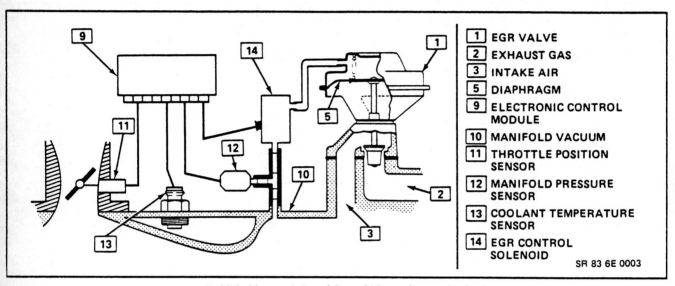

(Published by permission of General Motors Corporation.)

COMPUTERIZED EGR SYSTEMS

Due to the severe effects of EGR on engine performance, the EGR system became a prime candidate for microprocessor control. In the simplest cases, a computer-controlled solenoid takes the place of a temperature vacuum switch. In more complex applications, the vacuum signal is totally controlled by computer commands. Regardless of the application, the computer control of EGR has virtually eliminated the drivability problems inherent to exhaust-gas recirculation.

The computer monitors almost every engine operating condition through the sensory network. Prior to the computer, the amount of intake dilution was controlled only by ported vacuum signals, temperatuure vacuum switches, and exhaust backpressure signals. The computer is much more accurate because it can control EGR in relation to the following sensory inputs: engine speed (rpm), manifold pressure, barometric pressure (BARO), vehicle speed (mph), throttle position, engine temperature, and in some applications gear position. Obviously, the computer is capable of controlling the oxides of nitrogen while maintaining optimum engine performance.

The computer regulates EGR by energizing and de-energizing solenoids, which in turn control a vacuum signal to the EGR valve. When the ignition switch is turned on, 12 volts is applied to the solenoid windings. The ground for the windings is in the computer. The computer issues EGR commands by either grounding (energizing) or ungrounding (de-energizing) the solenoid windings. The commands are determined by the sensory inputs received by the computer.

Figure No. 142

EGR with One Solenoid

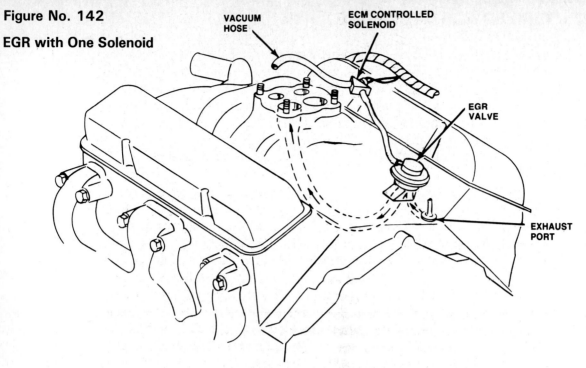

(Courtesy AC-Delco, General Motors.)

Figure No. 143

EGR with One Solenoid with Bleed Orifice

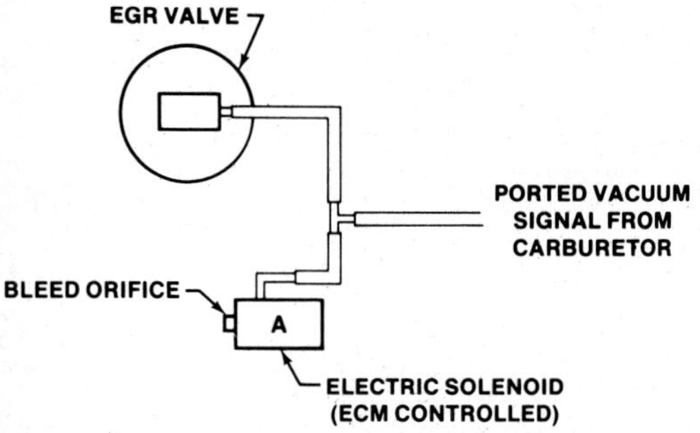

(Courtesy AC-Delco, General Motors.)

Figure No. 144

EGR with Two Solenoids

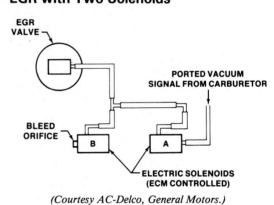

(Courtesy AC-Delco, General Motors.)

Figure No. 145

EGR with Auxiliary Vacuum Pump

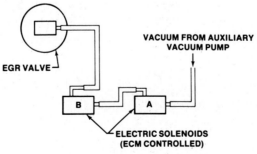

(Courtesy AC-Delco, General Motors.)

© Copyright 1986, Tune-Up Manufacturers Institute

GENERAL MOTORS EGR CONTROL SYSTEMS

Single-Solenoid EGR Control

A single solenoid is spliced in series with the ported vacuum signal line before the EGR valve (Figure No. 142). When the solenoid is energized by the computer, the ported vacuum signal is blocked and no EGR is permitted. The computer energizes the solenoid when the engine is cold, during cranking, and at wide-open throttle (WOT). The computer is constantly reading the coolant temperature sensor inputs and de-energizes the solenoid at a programmed temperature level. When de-energized, the solenoid is open, allowing ported vacuum and exhaust backpressure to regulate EGR flow.

Single Solenoid with Bleed Orifice

A single solenoid is fitted to the ported vacuum line, placing it in parallel with the EGR valve (Figure No. 143). Unlike the previous solenoid, the bleed orifice type does not block the ported vacuum signal. Instead, the solenoid will open the bleed orifice when energized by the computer on a cold engine. The open orifice reduces the ported vacuum signal, limiting EGR flow to less than half the warm-engine rate. After normal operating temperature is attained, the solenoid is de-energized and the bleed orifice is closed.

Dual-Solenoid Controls

Dual-solenoid controls are basically a combination of the single-solenoid and bleed-orifice-type systems (Figure No. 144). A blocking solenoid (A) is spliced in series with the EGR valve, and a bleed orifice solenoid (B) is placed in parallel with the valve.

The blocking solenoid takes the place of a temperature vacuum switch. During cold engine operation, the solenoid is energized, blocking the ported vacuum signal. The solenoid is open on a warm engine.

The bleed orifice solenoid reduces the ported vacuum signal when energized. The computer de-energizes the solenoid and allows full EGR flow when the following conditions are satisfied:

1. Closed-loop operation is attained.
2. Vehicle speed (mph) exceeds program calibration of the computer.
3. Run time calibration after hot start has elapsed.

All these conditions must be met before normal EGR operation is assumed.

Dual-Solenoid Controls with an Auxiliary Vacuum Pump

Some vehicles are equipped with an electric vacuum pump that generates the constant vacuum supply needed to operate engine controls and accessories (Figure No. 145). Since the pump delivers a constant vacuum, no throttle timed (ported) vacuum signal is present. Thus the computer must take over this control through a pair of solenoids. Both solenoids are blocking types and no bleed orifice type is used.

Solenoid A acts as the ported vacuum signal control. The computer de-energizes the solenoid, allowing vacuum flow when the following conditions are attained:

1. Closed-loop operation is attained.
2. Vehicle speed (mph) exceeds program calibration.
3. Run time calibration after hot start has elapsed.

Solenoid B acts as a TVS and load controller. Coolant temperature and manifold pressure inputs are monitored by the computer in order to determine command output. When manifold vacuum drops below calibration, and/or when engine temperature is below calibration, the solenoid is energized, thus blocking vacuum. High to moderate manifold vacuum and normal engine temperature allow the computer to de-energize the solenoid.

Notes

EGR SYSTEM USED WITH EEC-III

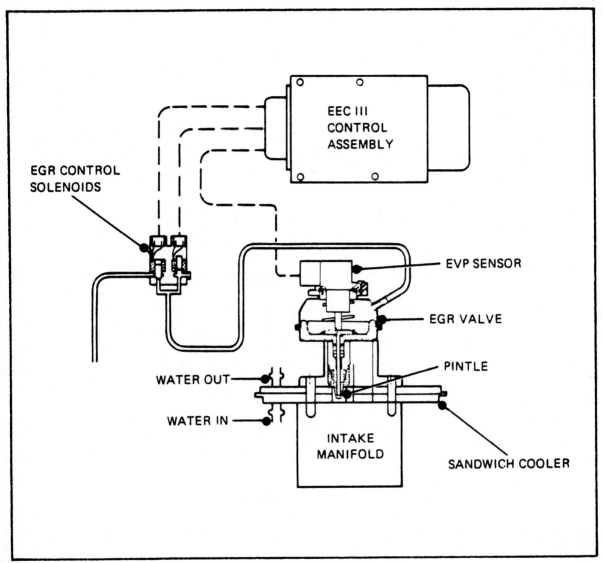

(Courtesy Ford Motor Company.)

LATE-MODEL FORD EGR SYSTEMS

Ford utilizes positive backpressure transduced EGR valves on most applications. The backpressure sensor may be either the remote style or integral. On integral applications, Ford uses a universal EGR valve for replacement of the original. The universal replacement valve comes with a bag of washers and an instruction sheet. The washers have different inside diameters. The instruction sheet determines which washer is to be inserted into the EGR valve flow port. The universal EGR valve allows dealers to stock one valve for many different applications and thus reduce the parts availability problems faced by mechanics.

Ford EGR with Electronic Control

On Fords with electronic engine control (EEC) systems, a unique form of computer control is used. The EGR valve is a single-diaphragm type and is *not* equipped with a backpressure sensor. The EGR pintle is tapered to control the volume of EGR flow in relation to valve stem position. The higher the diaphragm is lifted, the greater the EGR flow rate.

Unlike any other system, an EGR valve position (EVP) sensor is mounted atop the valve. This sensor is a variable resistor. The computer (ECA) sends a fixed reference voltage to the sensor. As the valve opens and closes, the resistance in the sensor changes, sending a varying voltage level to the computer through a signal wire. The computer is programmed to read the returning voltage as a specific EGR valve opening.

The computer monitors several engine operating conditions: coolant and intake air temperature, throttle position, manifold pressure, and engine speed. This information allows the computer to determine how much EGR flow is required. A pair of solenoids mounted on a valve cover are alternately energized and de-energized by the computer. These solenoids control the manifold vacuum applied to the EGR valve diaphragm. The EGR vent solenoid bleeds off vacuum to atmosphere when de-energized and traps vacuum in the EGR valve when energized. The EGR control solenoid is normally closed and opens when energized, allowing manifold vacuum through to the valve diaphragm. Computer control of the solenoids causes one of the following functions to occur:

1. *Increase* EGR flow by energizing both solenoids and applying vacuum to the EGR valve diaphragm.
2. *Decrease* EGR flow by de-energizing both solenoids and bleeding off the vacuum signal to the valve.
3. *Maintain* EGR flow by de-energizing the control solenoid (blocking manifold vacuum) and energizing the vent solenoid. This combination traps vacuum above the EGR valve diaphragm and holds it in a fixed position (flow rate).

During normal operation the solenoids are constantly cycling between the three functions. This rapid cycling is referred to as the *dithering* process. As with other EGR systems, the solenoids close the EGR valve during idle, cranking, wide-open throttle, and cold-engine operation.

Another feature peculiar to the Ford EEC system is the use of a cooler to reduce recirculated gas temperatures. The exhaust gases flow through a cooler (using engine coolant flow) that is sandwiched between the EGR valve and throttle body assembly. The cooler reduces combustion chamber temperatures with less intake charge dilution.

Testing of the system is too complex to cover in this text, but a general rule of thumb can be applied: if the solenoids can be felt or heard cycling, the system is operating.

Notes

Notes

TESTING EGR SYSTEMS

A malfunctioning EGR system creates numerous drivability and performance problems. Failures in this system can often be misleading, causing technicians to wrongly diagnose ignition or fuel systems. Hesitation, surge, poor idle, and spark knock are common symptoms of a defective EGR valve that are often incorrectly attributed to other engine systems. The following test procedures are basic to all EGR systems, but do not pertain to the idiosyncrasies of particular model applications. Use of a repair manual is required to properly test and correct EGR problems. Remember when working with components exposed to exhaust gases that extreme heat dangers exist. The wearing of gloves is a good safety precaution.

Single-Diaphragm EGR Valves

1. Disconnect the vacuum signal hose from the EGR valve and attach a hand-actuated vacuum pump to the valve. While applying and releasing vacuum, observe the valve-stem movement. Vacuum applied should raise the stem. Vacuum released should lower the stem. Failure to do so indicates a ruptured diaphragm or binding valve.
2. With a warm engine at idle, apply vacuum to the EGR valve. Engine speed should drop considerably or stalling should occur. This indicates that the valve is opening and that the EGR manifold passages are clear, allowing exhaust gases to enter the intake. If the opening of the valve does not affect idle quality, the passages are probably carbon-blocked and require cleaning.

Positive Backpressure EGR Valves

1. To check the EGR manifold passages, physically lift the valve stem and note idle quality. The loss of idle will be the same as it is for the single-diaphragm valve test.
2. Since the backpressure transducer controls the vacuum bleed port, vacuum applied to the EGR valve when the engine is off cannot lift the diaphragm. If the valve lifts, the transducer is clogged and the valve must be replaced.
3. A running engine check requires the exhaust to be partially blocked in order to create the backpressure needed to activate the sensor components. Securing a shop rag over the end of the tailpipe with a hose clamp is a good method. Place the throttle on the high step of the fast idle cam and apply vacuum to the EGR valve. A noticeable drop in rpm should take place.

Negative Backpressure EGR Valves

1. With a warm engine at idle, manually lift the EGR valve stem and take notice of idle quality. A deterioration of idle indicates clear passages of exhaust gases.

© Copyright 1986, Tune-Up Manufacturers Institute

2. Apply vacuum to the EGR valve with a hand-actuated vacuum pump while the engine is switched off. The valve should open and hold vacuum for approximately 15 seconds.
3. Applying vacuum to the valve while the engine is running should cause the valve to modulate.

Ported Vacuum Signal

1. Connect a vacuum gauge to the line between the carburetor and temperature vacuum switch.
2. Note the vacuum reading at idle. No vacuum should be present at idle. If vacuum exists, check idle adjustment or improperly routed lines.
3. Raise engine speed and observe gauge reading. The vacuum gauge should read approximately 10 inches of mercury.

Temperature Vacuum Switch

1. Connect a vacuum gauge between the EGR valve and the temperature vacuum switch (TVS).
2. On a cold engine, raise the engine speed to approximately 2000 rpm and observe the gauge reading. Regardless of throttle opening, no vacuum should be present. If a vacuum signal appears, replace the TVS.
3. Repeat this test on a warm engine. A vacuum signal should be present at off idle.

Notes

FUEL EVAPORATION EMISSION CONTROL SYSTEMS

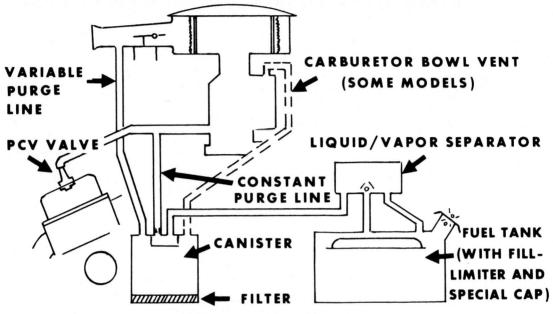

CANISTER STORAGE TYPE

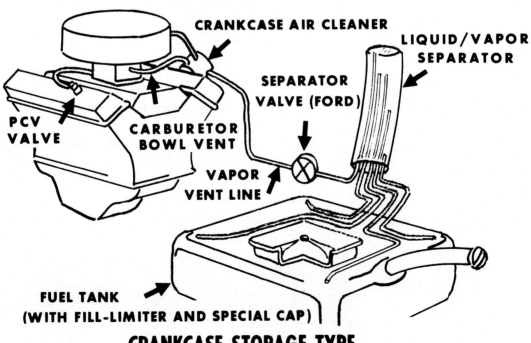

CRANKCASE STORAGE TYPE

FUEL EVAPORATION EMISSION CONTROL SYSTEMS

The fuel evaporation emission control system was developed and designed to prevent the escape of gasoline vapors from the fuel tank and carburetor into the atmosphere. This system is in addition to the positive crankcase ventilation system and the various exhaust emission control systems.

Starting with the 1972 car models, all vehicles use the system of storing the fuel vapors in a charcoal-granule-filled canister. Some earlier models also used this system, while others employed the engine crankcase as the storage area for the fuel vapors.

Essentially, the evaporation emission control systems function in the following manner: When the engine is running, the fuel vapors are conducted through tubing from the fuel tank to the carburetor or carburetor air cleaner to be immediately consumed in the engine. When the engine is not running, the tank and carburetor fuel vapors are either piped into a charcoal-granule-filled canister or they are stored in the engine crankcase, depending on system design. When the engine is started, air passing through the canister, or through the crankcase ventilation system, picks up the fuel vapors and carries them into the engine where they are consumed.

By this method of control, fuel vapors are contained in the evaporation-control-system, thereby eliminating fuel vapors as a source of air pollution.

Vapor Transfer System

The fuel tank cap is either nonvented or is a pressure-vacuum sensitive type, thereby preventing the escape of either liquid gasoline or gasoline vapors through the tank filler. To assure that the tank will be constantly vented, two, three, or four points are used as tank vents. Lines from these tank-vent points meet at a unit called a liquid-vapor separator, which is positioned just forward of, and slightly above, the fuel tank. Only fuel vapors can pass the separator and enter the single vapor transfer line leading forward to the canister or crankcase. Liquid fuel is returned to the tank through a return vent line.

The separator front vent section is equipped with a needle-and-seat (or ball-and-seat) positive shut-off valve. In the event of vehicle upset, liquid gasoline cannot flow uncontrolled into the canister or crankcase. Because of fuel tank position on some station wagons, a separator is not used on these vehicles.

On some vapor-control systems, a two-way pressure-vacuum relief is positioned just forward of the separator. The valve's function is to prevent fuel starvation or mechanical damage to the fuel tank. If fuel tank pressure falls too low or rises too high, the valve will admit air or expel vapor through the transfer line, thereby restoring normal pressure values. On some systems the valve maintains a positive pressure of 1 pound per square inch in the fuel tank and vent system to aid in retarding further fuel evaporation.

Charcoal Canister Storage

From the relief valve (or separator) the vapor transfer line runs to an activated charcoal-containing canister located in or near the engine compartment. Another line leads from the canister to the carburetor PCV hose. As previously stated, when the engine is running, the tank fuel vapors are constantly being drawn into the engine. When the engine is shut down, the fuel vapors generated by "heat soak" are absorbed into the activated charcoal. When the engine is restarted, engine vacuum draws air through the canister filter over the charcoal granules, removing the absorbed fuel vapors and drawing them into the engine.

Some vapor control systems employ two vent lines from the canister. One line runs from the canister to the carburetor PCV hose, previously mentioned. This line is called the *constant flow purge line.* Its function is to permit the engine manifold vacuum to draw a constant uniform amount of fuel vapor from the canister at all engine speeds and operating conditions, including engine idle periods. Vapor flow is controlled by a fixed orifice positioned in the top of the canister. The second line runs from the canister to the carburetor air cleaner snorkel. This line is usually called the *variable flow purge line.* The vapors that pass into the engine through this line are in variable amounts depending on engine speed, particularly at speeds above idle. This arrangement permits complete canister purge at cruising speeds.

The activated charcoal granules contained in the canister have a capacity of approximately 50 grams of vapor, which is the equivalent of between 2 and 3 ounces of liquid gasoline. This capacity is adequate to effectively contain fuel vapors even when the vehicle is parked for an extended period.

Crankcase Storage

All 1970-1971 engines built by Chrysler Corporation and some Ford-built engines of the same years used the engine crankcase as the fuel vapor storage area. Since fuel vapors are from two to four times heavier than air, they settle in the crankcase on top of the oil. When the engine is running, the fuel vapors are purged from the crankcase and drawn through the crankcase ventilation system into the engine. On the Ford engine a combination valve placed in the vapor line serves to isolate the fuel tank from engine-induced pressures.

Carburetor Vapor Control

The control of carburetor fuel vapor is particularly effective during periods of "heat soak," which occur immediately after engine shutdown. These are the periods of maximum carburetor fuel vapor emission.

The method of carburetor vapor control consists of a line from the carburetor bowl to the crankcase on some installations or, on some four-barrel carburetors, a line connecting the carburetor primary fuel bowl to the vapor canister.

In other installations, the generation of fuel vapors is somewhat minimized by the use of an aluminum heat-dissipating plate, which contains the carburetor to intake manifold gasket. The plate serves as a heat shield to deflect and dissipate the engine heat to the surrounding air, thereby maintaining a relatively low carburetor fuel temperature. The possibility of fuel boil-away is consequently considerably relieved.

Fuel Tank

A slight redesigning of the upper portion of the fuel tank is also a part of the vapor emission control system. An overfill protector or fill-limiter (an inverted dishlike member) is mounted in the top of the fuel tank. In some installations this member is actually a small separate tank with a capacity of slightly more than 1 gallon and is connected to the main tank by small metered holes. Regardless of its size or shape, the limiter is designed to remain essentially empty after the fuel tank is filled. A fuel tank top area collection space is thereby maintained for the fuel vapors and an expansion space for increased fuel volume. These conditions occur when a filled fuel tank is subjected to high temperature, as when the vehicle is parked in the hot sun.

Service

Service of the fuel evaporation emission control system is generally limited to cleaning or replacing the canister filter at certain intervals. The separator is a sealed unit and requires no service or maintainence.

When an engine equipped with a vapor control system is being tuned, the fuel tank vent line should be disconnected at the canister. The line should not be plugged. Disconnecting the line prevents fuel vapors from being drawn into the engine and upsetting final adjustments.

Figure No. 148
Canister Control Valve

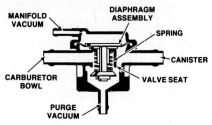

(Courtesy AC-Delco, General Motors.)

Figure No. 149
Canister Control Valve

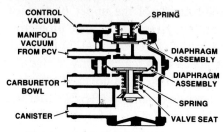

(Courtesy AC-Delco, General Motors.)

Figure No. 150
Canister — Vapor Vent Valve Type

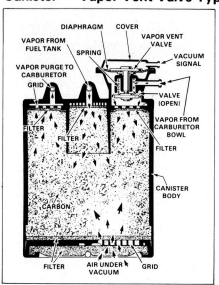

(Courtesy AC-Delco, General Motors.)

Figure No. 151
Canister — Vapor Vent Valve and Purge Valve Type

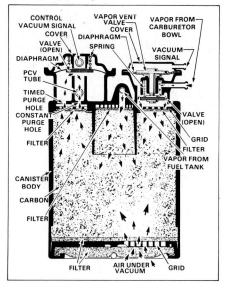

(Courtesy AC-Delco, General Motors.)

Figure No. 152

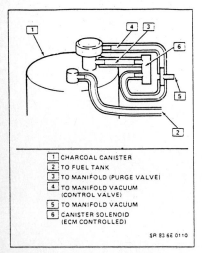

(Courtesy AC-Delco, General Motors.)

Figure No. 153

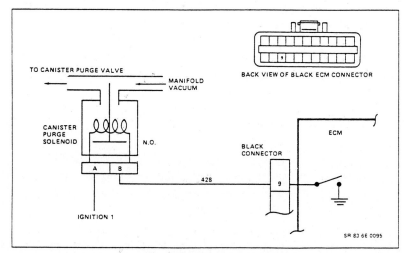

(Courtesy AC-Delco, General Motors.)

LATE-MODEL PURGE CONTROL

Ever tightening emissions standards have caused the vehicle manufacturer to more precisely control evaporative pollutants. Prior to 1980, most cars relied on a constant purge of the vapor canister and did not pay much attention to carburetor bowl venting. With the advent of computerized engine controls, the introduction of fuel vapors would have detrimental effects on the critical balance maintained by the closed-loop system. These hydrocarbons could cause the system to react incorrectly and create drivability problems. Therefore, the timing of purge had to be strictly controlled.

Bowl Vent Controls

There are two types of bowl vent controls: vacuum and electrically operated. Regardless of the type used, one basic rule applies to late-model bowl venting systems. The bowl vent must be closed when the engine is running and open when the engine is switched off.

Electric solenoid controls are the most common form of bowl vent control. This solenoid may be located in the vent line to the canister or constructed as an integral part of the carburetor air horn assembly. Bowl vent solenoids are *not* controlled by the computer. When the ignition is switched on, the solenoid energizes and blocks the flow of bowl vapors to the canister. Turning off the ignition de-energizes the solenoid, allowing hot soak vapors to enter the canister instead of flooding the intake.

The vacuum-operated bowl vent valve may be located in the vent line or on the canister. The valve has three hose connections: canister line, bowl vent line, and manifold vacuum signal line. The manifold vacuum signal generated by a running engine acts against a diaphragm, which closes the passage of vapors to the canister. When the engine is switched off, the valve opens and bowl venting is allowed. Some vacuum-operated valves also contain bimetal thermal discs to control venting in relation to underhood temperatures. Whether thermal control is used or not, the vacuum signal is the primary control and overrides the bimetal discs, closing off the vent line anytime the engine is running.

Should the bowl vent remain open on a running engine, a rich-mixture condition would result. This richness would affect the oxygen sensor signal to the computer. The computer cannot sufficiently correct the problem and catalytic converter meltdown is a common occurrence. Logic-type computer systems (computers with memory) set trouble codes indicating "rich oxygen sensor" when exhaust oxygen content is constantly low (rich exhaust). Do *not* mistakenly condemn the oxygen sensor. A malfunctioning bowl vent valve is the most common cause of this condition.

Computer-Controlled Purge

As is the case with other computer controls, a solenoid is used to regulate purge of the vapor canister. The ground for the solenoid windings is controlled by the computer. By energizing or de-energizing the solenoid, vacuum to the canister can be applied or withdrawn. On General Motors vehicles, the solenoid is energized to prevent canister purge and de-energized to open the vacuum line to the canister. Ford reverses the procedure by blocking vacuum when de-energized and opening when energized.

© Copyright 1986, Tune-Up Manufacturers Institute

The computer determines when purge is to occur by consulting the following sensors: coolant temperature, throttle position, vehicle speed, and elapsed running time. A typical control sequence can be best illustrated by the following General Motors example:

The following conditions must be met before the computer (ECM) will de-energize the purge control solenoid and allow canister purge:

1. Coolant temperature above 75°C.
2. Throttle off idle.
3. Vehicle speed above 15 mph.
4. Engine run time after start more than 1 minute.

Notes

Figure No. 154

MANIFOLD HEAT CONTROL VALVE

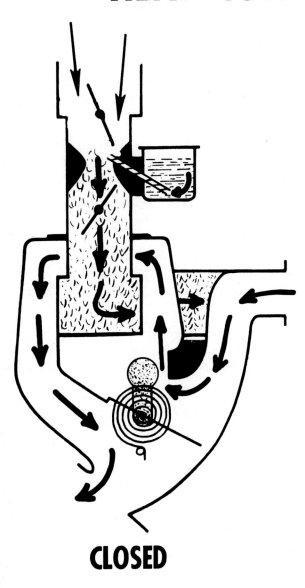

CLOSED

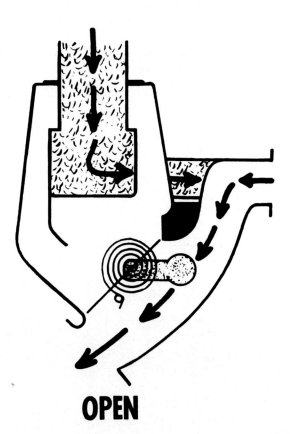

OPEN

MANIFOLD HEAT CONTROL VALVE (HEAT RISER)

The intake manifold depends on heat from the exhaust manifold to assist in proper fuel vaporization. A thermostatically controlled valve arrangement provides the correct heat transfer between the exhaust manifold and the intake manifold. Most models use a thermostatic spring in the automatic choke, which is also actuated by heat from the exhaust manifold. Some late-model engines (1974 and up) use a vacuum-actuated coolant-temperature-controlled manifold heat-control valve. This system is called an EFE (early fuel evaporation) by GM and VHC (vacuum-operated heat control) by Ford. If intake manifold temperatures do not rise in proportion to the exhaust manifold temperatures, an unbalanced condition between the automatic choke and the intake manifold temperature will result. For instance, if the intake manifold is retarded in reaching operating temperature due to improper heat control valve operation, the engine will have the effect of running lean because of insufficient vaporization of the fuel. Meanwhile, the exhaust has caused the automatic choke to open, providing a still leaner mixture.

When the engine exhaust manifold is cold, the heat control valve is held in a closed or *heat on* position. In this position the exhaust gas is deflected upward and around the intake manifold hot spot, then downward into the exhaust pipe. Once the intake manifold has reached its normal operating temperature, the heat-control valve is moved to the open or *heat off* position, and the exhaust gas is deflected directly from the exhaust manifold into the exhaust pipe.

If the manifold heat control valve sticks in the open position, engine warm-up is delayed because the desired preheating action of the fuel charge does not take place. This means the choke stays in operation for a long time, increasing fuel consumption and diluting the motor oil.

If the heat valve sticks in the closed position, the fuel charge is constantly subjected to preheating even after engine temperature is normalized. This condition results in a lean fuel mixture, loss of power, hard "hot engine" starts, and possible burning of the engine valves.

The valve and mechanism should be checked for free operation as necessary. It should move freely approximately 90 degrees without evidence of sticking or binding. If there is a tendency of the valve to stick, it should be hand operated while the recommended lubricant (hot motor oil) is liberally applied to the valve shaft ends. If, due to obstructions, the lubricant cannot be applied directly to the valve shaft ends, it is often possible to touch a long screwdriver to each shaft end and let the lubricant run down the screwdriver shank.

If the valve cannot be moved by hand, lightly tap the shaft ends with a small hammer to break the shaft loose. Then apply the lubricant while hand operating the valve. The actuator should be checked with a vacuum source for movement and leakage. Check TVS for proper operation, and inspect all hoses for cracks and the like. Replace any parts if necessary.

© Copyright 1986, Tune-Up Manufacturers Institute

Figure No. 155

ELECTRIC GRID EFE

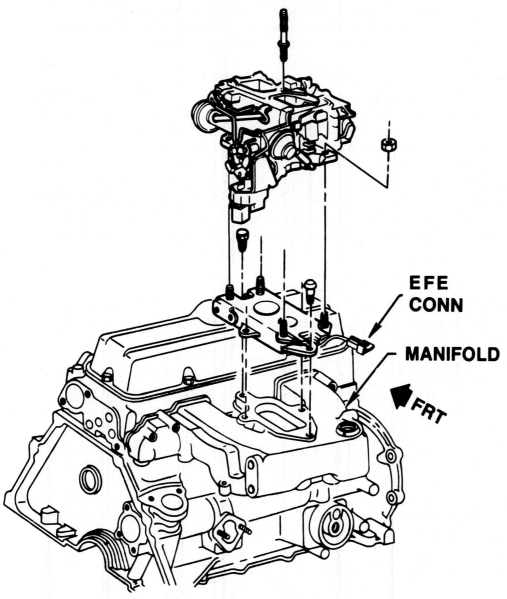

(Courtesy AC-Delco, General Motors.)

ELECTRIC GRID EARLY FUEL EVAPORATION (EFE)

While most other engine control devices have grown in complexity, here is one that has actually been simplified. Heat risers and vacuum-controlled EFE systems regulate exhaust gas flow under the intake manifold, which reduces exhaust efficiency and increases the chances of malfunction. The electric grid EFE uses a simple ceramic heating element located under the primary bores of the carburetor. The ceramic grid is an integral part of the insulator gasket between the carburetor and intake manifold. The grid heats the incoming intake charge in order to improve vaporization and cold engine drivability.

There are two types of electric EFE grids, single- and double-wire units. The double-wire type is controlled by a thermal switch located in the water jacket. When coolant temperature is below a calibrated level, the switch closes and current is supplied to the heating element. When temperature rises above calibration, the switch opens, blocking the flow of current to the EFE grid.

Single-wire units are controlled by an on-board computer. The computer continually monitors engine temperature by analyzing data received from the coolant temperature sensor. When the engine is cold, the computer arms the EFE relay, sending current to the EFE grid. On a warm engine, the computer opens the relay, shutting off the EFE. Single-wire units usually have a ground wire connected to the insulator gasket.

Testing Electric EFE Systems

To test the electric EFE, a 12-volt test light and ohmmeter are required.

1. Connect the test light to the leads going to the grid. The engine should be cold.
2. The light should glow brightly when the ignition is switched on. If it does not, check for an open or short in the EFE circuitry.
3. Connect the ohmmeter across the EFE grid connectors. The resistance should be less than 3 ohms. If not, replace the grid.

It is important that objects not come in contact with the heating grid. Avoid the practice of inserting a screwdriver down the carburetor bore to hold the throttles open during compression tests. The screwdriver can damage the element or create a hazardous short.

POSITIVE CRANKCASE VENTILATION SYSTEMS

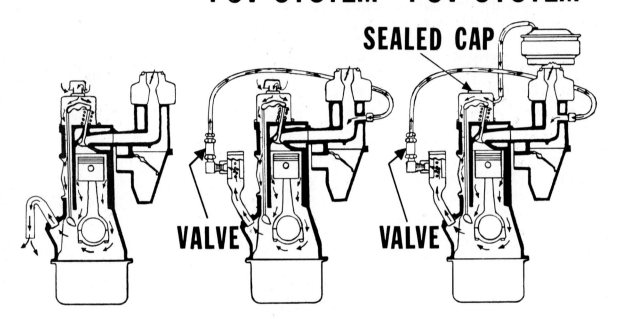

PCV METERING VALVE

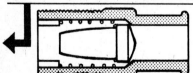

ENGINE STOPPED
(VALVE CLOSED)

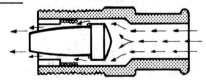

IDLING OR LOW-SPEED AIR FLOW RESTRICTED
(VACUUM HIGH)

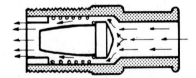

HIGH SPEED MAXIMUM AIR FLOW
(VACUUM LOW)

© Copyright 1986, Tune-Up Manufacturers Institute

POSITIVE CRANKCASE VENTILATING SYSTEMS

During engine operation certain amounts of fuel charge and exhaust gases find their way past the piston rings into the crankcase. These gases are commonly called *blow-by*.

The presence of blow-by gases and moisture condensation in the crankcase causes contamination of the motor oil. This combination of elements results in sludge formation in the oil. In a neglected state this sludge can block the oil pump screen and oil galleries with disastrous results. Further, these contaminants are highly corrosive and will eventually result in acid etching and rusting of highly polished internal engine surfaces. Varnish and lacquer formation on critical engine parts, such as hydraulic lifters and cam lobes, will seriously interfere with efficient engine operation. Proper ventilation of the crankcase to remove these fumes, gases, and condensed vapors is essential to prevent motor oil contamination and extend engine life.

Draft Tube

Ventilating the crankcase by passing air thorugh it has been the standard practice for many years. Air entered the crankcase through the oil fill cap, or crankcase breather as it was frequently called, passed through the crankcase picking up the fumes and gases, and exhausted from the engine through a road draft tube.

This system had two disadvantages. First, the amount of air passing through the crankcase is governed by how fast the car is being driven. The movement of air under the vehicle is the factor that created a greater or lesser pressure differential at the end of the draft tube with the air pressure at the breather cap. At idle and low cruising speeds, the passage of air through the crankcase is inadequate for effective ventilation. Second, the fumes ventilated from the crankcase are exhausted into the atmosphere, thereby contributing to air pollution.

Open Positive Crankcase Ventilating System

The positive crankcase ventilating system, frequently referred to as PCV, assists in reducing the air pollution caused by automobile crankcase vapors. In the PCV system the draft tube is eliminated. Blow-by gases and vapors are drawn from the crankcase by intake manifold vacuum, through tubing into the intake manifold, and into the engine cylinders where they are burned. The rate of air flow through the system is controlled by a valve. A few systems employ a fixed metered orifice instead of a valve. A slight valve movement meters the air flow to provide the proper degree of crankcase ventilation consistent with engine demands and vehicle speeds.

With the elimination of the major portion of the blow-by gases, engine oil contamination is reduced, water vapor, rust, and corrosive elements are largely eliminated, and engine life is materially extended. All these benefits are secured while air pollution is considerably reduced.

When service of the PCV system is neglected, the valve usually ceases to function, since it is in the line carrying the crankcase contaminants. When this happens, the system is blocked, crankcase pressure develops, and the crankcase fumes and blow-by gases are forced out of the crankcase backward through the breather cap. Not only are the benefits of the PCV system lost but the crankcase fumes are again polluting the atmosphere. This undesirable factor led to the development of the *closed* PCV system.

Closed PCV System

In the closed PCV system the crankcase breather cap is replaced with a solid-type cap similar to a gas tank cap. Air entering the system is introduced through the carburetor air cleaner. Air flow through the system is controlled by the action of a valve influenced by manifold vacuum, the same as in the standard PCV system. However, should the closed system be neglected to the extent that crankcase pressure causes a reverse flow of blow-by gases and fumes, they will be drawn into the carburetor through the air cleaner. The smog-producing fumes will not be vented into the atmosphere but will be burned in the engine.

The closed PCV system was made mandatory by federal law and has been standard equipment on all passenger car and light truck engines starting with the 1968 models.

Testing and servicing procedures for the PCV system are covered in the tune-up procedure, Chapter 12.

Notes

Figure No. 157

POSITIVE CRANKCASE VENTILATION SYSTEM TESTS

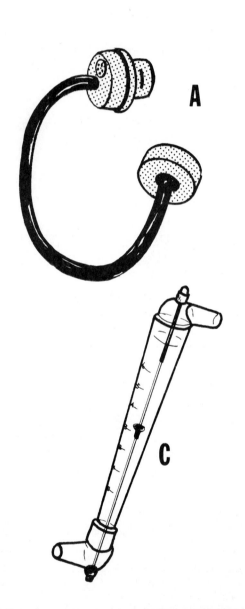
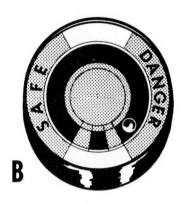
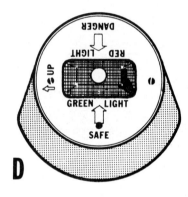
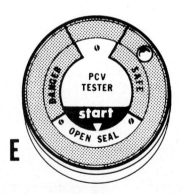

VARIOUS TYPE TESTERS

POSITIVE CRANKCASE VENTILATION SYSTEM TESTS

Several gauges are marketed for testing the operation of the crankcase ventilating system. The tests performed are essentially checks on the operation of the PCV valve. The testers are all basically pressure-sensing devices. Their function is to check the degree of air flow and flow-by circulation through the crankcase while the engine is idling.

One tester (Figure No. 157A) is equipped with a selection or rubber adapters to plug engine breather openings, such as the oil dipstick tube or the dual breather cap used on some V-8 engines. A selector knob on the bottom of the tester is adjusted to match the air flow rate of the valve on the engine being tested. With the engine idling, and at normal operating temperature, the breather cap is removed from the rocker arm cover, and the tester adapter is pressed into the breather cap opening. With the tester held upright, the color indicator can be viewed through the tester window. The green color indicates proper air flow circulation so the PCV system is functioning properly. A yellow color means the system is partially plugged or the crankcase is not properly sealed. The red color indicates a blocked system, which generally means a plugged valve. In the case of a worn engine, blow-by past the rings may be so excessive that it is beyond the capacity of even a properly operating PCV system to handle. In this event the engine cannot be tuned with any degree of success.

Another tester (Figure No. 157C) differs in design from the other gauges. A hose connected to the PCV systems is attached to the bottom of the tester. A hose leading to the carburetor or intake manifold is attached to the top of the tester. A free-sliding indicator is mounted on a rod housed inside the tester. The degree of vacuum passing through the system is indicated by the reading on the side of the tester body opposite the position of the sliding indicator.

The other testers are variations of the same principle. The crankcase breather is removed and the tester is firmly seated in the breather cap opening. Vacuum or pressure in the system casues the ball in the tester to indicate a safe or danger position. One of the testers (Figure No. 157E) also indicates an *open-seal* position, which warns of possible leaks in the PCV system.

Full operating instructions are included with every tester.

When tests and service of the PCV system are not performed at every tune-up, there are several undesirable effects other than air pollution.

1. Rough engine idle. A plugged or restricted PCV system upsets the carburetor air-fuel ratio. With the idle mixture unbalanced, a rough idle and a tendency to stall condition exist.

2. Increased fuel consumption. Because of the unbalanced air-fuel ratio, there is a proportionate increase in fuel consumption. The reduction in air supply will produce a partial-choke condition.

3. Loss of engine performance. Spark-plug gas fouling associated with rich carburetor mixtures is quickly reflected in loss of acceleration, misfiring, and generally poor engine performance.

4. Increased oil consumption. Crankcase pressure developed by a blocked system can force oil past front and rear main bearing seals, past rocker arm and oil pan gaskets, and out the crankcase dipstick tube. A PCV valve stuck in the

open position can cause oil to be pulled out of the oil pan into the engine. This condition will result in a depleted oil supply and a ruined engine.

5. Motor oil sludge buildup. Unvented moisture vapors and blow-by gases condense and settle in the motor oil, resulting in sludge formation robbing the engine of its vital lubrication.
6. Premature engine wear. Motor oil contamination destroys the additives in the motor oil, causing acid etching and rusting of critical high-polished internal engine surfaces. This condition results in accelerated engine wear. Cam lobes and hydraulic valve lifters are particularly affected by this condition.
7. Crankcase odors. Fumes trapped in the crankcase give off a strong odor of hot oil that often finds its way into the car. These odors are sickening and nauseating to most people. Crankcase odor in the car is one of the first indications of a neglected PCV system.

No professional tune-up can be performed without including the testing and servicing of the PCV system and replacing the valve as required.

Notes

Figure No. 158

CATALYTIC CONVERTER — PELLET TYPE

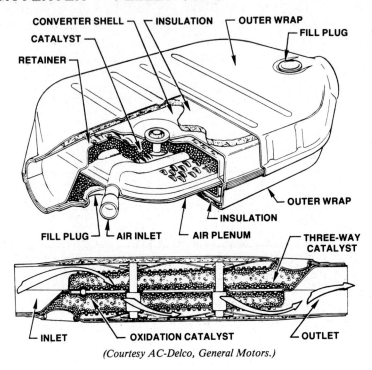

(Courtesy AC-Delco, General Motors.)

Figure No. 159

CATALYTIC CONVERTER — MONOLITH TYPE

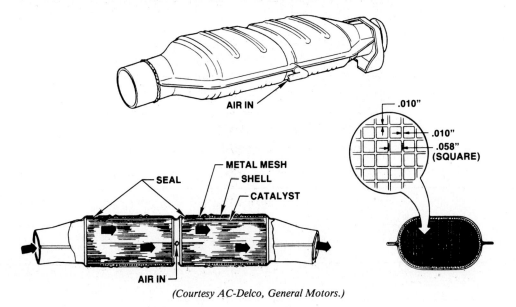

(Courtesy AC-Delco, General Motors.)

CATALYTIC CONVERTERS

As legislated emission requirements tightened, vehicle manufacturers realized that tuning modifications and add-on components were not sufficient. The use of an exhaust-gas reaction device located in the exhaust flow was the obvious solution to excessive pollutants. A material is considered to be a catalyst when it alters the rate of a chemical reaction or guides it in a specific direction without itself being affected. The material selected was platinum (or synthetic platinum derivatives), and its enclosure in a mufflerlike container was called the *catalytic converter*.

Oxidation Catalysts

An oxidation catalyst is often referred to as a conventional type because it was the first catalyst used and dominated the industry during the 1970s. There are two types of converter construction, pellet and monolith.

The pellet type (Figure No. 158) consists of a baffled container filled with aluminum oxide pellets. As the exhaust gases pass through the pellets, they come in contact with the catalytic materials. The pellets can often be changed without replacing the converter housing.

The monolithic type (Figure No. 159) is constructed of a single honeycomb element coated with platinum and palladium. At the inlet to the converter, a diffuser plate is used to disperse the exhaust flow. This diffusion allows maximum use of the catalyst-material surface area. Monolithics are considered superior to pellets due to reduced backpressure, greater longevity, and more rapid heating cycle.

The treatment of exhaust gases with catalysts is a chemical process that generates extreme heat. When the platinum and palladium reach a sufficient temperature (500° to 600°F), the converter is said to *light off,* beginning the oxidation process. Carbon monoxide (CO) is combined with free atoms of oxygen (O_2) to create carbon dioxide (CO_2). Hydrocarbons (HC) are separated into hydrogen (H_2) and carbon (C) and combined with oxygen (O_2) to yield water (H_2O) and carbon dioxide (CO_2). The oxidation process is represented by the following chemical expression:

$$\text{Hydrocarbons} \quad HC + O_2 \rightarrow CO_2 + H_2O$$
$$\text{Carbon Dioxide} \quad CO + O_2 \rightarrow CO_2$$

This chemical reaction explains why two-gas emission analyzers cannot be used for tuning catalyst-equipped engines. A properly operating catalyst will mask CO and HC levels, requiring four-gas analyzers capable of measuring carbon dioxide and oxygen output.

Three-Way Catalyst

The three-way catalyst (TWC) contains oxidation materials combined with another rare metal, rhodium. Rhodium is a reduction catalytic material designed to reduce oxides of nitrogen. The TWC is capable of handling hydrocarbons, carbon monoxide, and oxides of nitrogen; thus the name three-way catalyst.

Upon converter light-off, the rhodium reduction process begins. As the oxides of nitrogen pass over the catalyst, they are separated into uncombined nitrogen (N_2) and carbon dioxide. The reduction reaction is best illustrated by the following chemical representation:

$$NO_x + CO \rightarrow N_2 + CO_2$$

Three-way catalysts come in two configurations. The reduction and oxidation materials may be enclosed together in a single converter housing. The materials may also be separated into two separate housings, with the rhodium unit located upstream of the oxidation converter.

Dual-Stage Converters

Dual-stage converters consist of a three-way catalyst and an oxidation catalyst located in a single converter housing. As exhaust gases pass through the TWC portion of the converter, most of the free oxygen is consumed. To enable the oxidation monolith, which is located after the TWC, to continue the treatment of HC and CO, additional air must be supplied to the converter. For this reason, an air space separates the two converter sections. A tube connects the air space to the air-management system, which supplies the additional oxygen needed in the second stage of conversion.

Testing

A four-gas emission analyzer is needed to test the conversion capabilities of the catalyst. While these testers are expensive and rarely found in garages, state and federal inspection laws will be requiring the purchase of four-gas equipment.

A common problem does afflict many converter-equipped cars, however. When more HC and CO are emitted than the converter can handle, the catalytic material becomes superheated and melts down. The result of the meltdown is exhaust restriction. Since the engine is an air pump, it can only take in as much air as it can put out. A restrictive converter will cause drastic power loss and can obstruct exhaust flow to the point of creating a no-start condition.

A simple vacuum gauge can be used to check for a restriction. Connect the gauge to a good manifold vacuum source on a fully warmed engine. Raise the engine speed to a constant rpm of between 2000 and 2500 and hold it there for approximately 30 seconds. Observe the vacuum reading. A normal vacuum will remain steady at 16 to 20 inches of mercury. Should the vacuum level decline during the test, an exhaust blockage is most likely the problem.

The causes of catalyst failure can vary. Any problem on the engine that creates exaggerated carbon monoxide (CO) and/or hydrocarbon (HC) output can cause converter meltdown. Secondary ignition component failure is the most common cause, but carburetor and purge system problems can be equally destructive.

The use of leaded fuel renders a converter inactive, but will not cause meltdown. The lead in the fuel merely coats the catalytic material, insulating it from contact with the pollutants. Remember, converters do not fail by themselves. If the converter is damaged, find the problem that caused the failure. There is no sense in replacing a catalyst only to have the problem return a few days later. Catalytic converters are very expensive.

Notes

Figure No. 160 EXHAUST GAS OXYGEN (EGO) SENSOR INSTALLATION

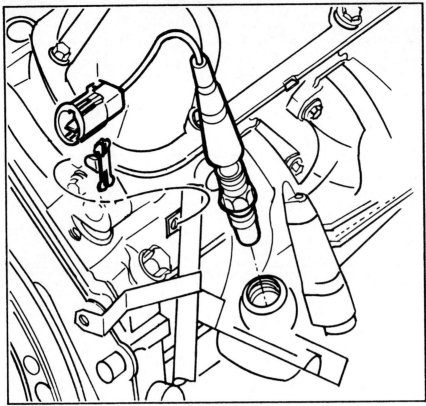

(Courtesy Ford Motor Company.)

Figure No. 161 EGO SENSOR

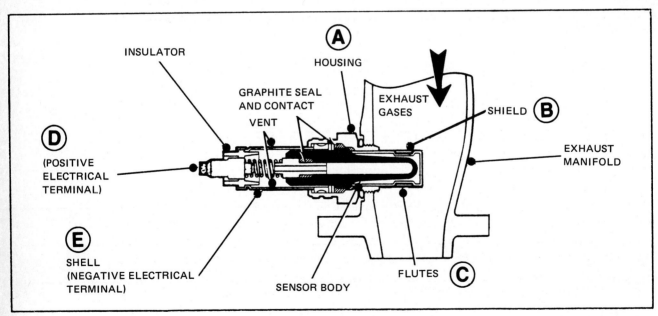

(Courtesy Ford Motor Company.)

OXYGEN SENSOR

During the 1970s, domestic manufacturers concentrated on the development of superlean-burning engines. To ignite these mixtures, special ignition systems (Chrysler Lean Burn, G. M. Mizar) were fitted to the engines. Spark-plug gaps were widened, and electronic ignition systems were strained to the maximum of their reserve capacities. While these lean air-fuel ratios satisfied EPA requirements, performance, drivability, and fuel economy suffered greatly.

European engineers pursued a different solution to emissions. They discovered that lean mixtures were not the answer. Their research found that there was a perfect mixture, allowing an emissions level that could easily be handled by a three-way catalytic converter, while maintaining near optimum performance and fuel economy. The Europeans referred to the ideal mixture as lambda; Americans call it a stoichiometric ratio. This mixture equates to an air-to-fuel *ratio* of approximately 14.7:1 by weight. The oxygen sensor was designed to monitor the mixture for a microprocessor, which in turn controls the fuel supplied by a carburetor or fuel injector(s), maintaining the stoichiometric ratio. The oxygen sensor (lambda probe) first appeared on Volvo cars in 1975, and by 1980 virtually every car sold in America was equipped with an oxygen sensor and three-way catalytic converter.

The oxygen sensor is fitted into the exhaust header close to the cylinder head and upstream of the catalytic converter (Figure 160). Because combustion is never totally complete, a certain amount of uncombined oxygen is always present in the exhaust gases. The amount of oxygen content present in the exhaust will vary with the initial air-fuel mixture strength. A rich initial air-fuel mixture creates a low exhaust oxygen content, while a lean mixture yields high exhaust oxygen levels. Since the sensor is exposed to the exhaust flow, it will be able to read the changes in oxygen content.

The external configuration of the O_2 sensor is similar to that of a spark plug, but its composition is more related to a catalytic converter (Figure No. 161). The sensing element within the sensor is constructed of a ceramic body coated with a gas-permeable platinum layer (zirconium dioxide). The outer surface of the element is exposed to exhaust-gas oxygen. The inner surface is vented to atmosphere and is always in contact with ambient oxygen. When heated to a temperature of 600°F by the exhaust gases, the difference in oxygen content between the inner and outer surfaces will generate a small voltage signal. Since a rich mixture yields a low oxygen content in the exhaust, the difference in oxygen levels between outer and inner surfaces of the sensing element is greatest, generating a voltage signal of 500 to 900 millivolts (0.5 to 0.9 volts). A lean mixture creates a higher oxygen level; thus the difference between inner and outer surface contact is reduced. A lean mixture generates a lower voltage signal, 100 to 500 millivolts (0.1 to 0.5 volts).

The oxygen sensor voltage signal is sent to the computer, which is programmed to read the varying voltage signal as a specific mixture value. The computer then calculates the length of time (duration, pulse width) a mixture-control solenoid or fuel-injector solenoid will be energized. The oxygen sensor, computer, and mixture solenoid(s) are referred to as the *loop*. It is important to understand the meaning of open-loop and closed-loop operation. When the oxygen sensor is being used by the computer to control mixture, the system is said to be operating in closed loop. Closed-loop operation does *not* mean that a constant 14.7:1 air-fuel ratio is maintained. The mixture strength will constantly vary between rich and lean, with an average ratio of 14.7:1 being the end result. For example, on a properly operating engine, the oxygen sensor reading will vary between 200 and 800 millivolts. Therefore, the computer will issue a solenoid duration command that varies from rich to lean.

When the computer is *not* using the oxygen sensor to control mixture, the engine is operating in open loop. During open-loop operation, the computer will maintain a constant fuel-delivery program and no variation in solenoid duration will exist. A system will enter open loop for several reasons. A cold oxygen sensor does not provide a voltage signal; thus the computer will not enter closed loop with a cold engine. The computer may also be programmed to enter open loop under certain operating conditions: wide-open throttle (WOT), idle, or deceleration. Of course, this programming will be different depending on the make and model of car.

The oxygen sensor is a delicate component and special precautions must be taken when servicing.

1. Never connect an ohmmeter to the sensor leads.
2. Never ground the sensor signal lead.
3. Always coat the threads with a glass and graphite antiseize compound before reinstalling a sensor.
4. Do not torque sensor beyond 30 pound-foot.
5. Do not spray chemicals in the area of the sensor. Contamination of the inner surface may occur.
6. Leaded fuel cannot be used in an oxygen-sensor-equipped vehicle. Lead coats the element, rendering it inactive.
7. Make sure gasket sealing materials used on engine covers are approved for automobiles with oxygen sensors.
8. Sensors should be replaced at approximately 30,000-mile intervals.

Testing

Testing an oxygen sensor requires the use of a digital volt-ohmmeter (DVOM) with minimum impedence of 10 megohms. Select the lowest voltage scale and connect the voltmeter between the sensor signal lead and ground. Ensure that the engine is up to normal operating temperature and observe the voltage readings. The meter should fluctuate evenly to each side of 500 millivolts. Create a lean condition by disconnecting a vacuum hose, and the output should fall below 500 millivolts. Create a rich mixture by slightly closing the choke valve, and the meter should read above 500 millivolts. Remember that if the engine remains at idle for a long period of time the sensor element will cool to below 600°F. If this happens, the sensor will not generate voltage and will test defective. Should sensor cooling be suspected, raise the engine speed to approximately 2000 rpm for 30 seconds in order to heat the element.

Notes

Figure No. 162

TRANSMISSION REGULATED SPARK (TRS) SYSTEM

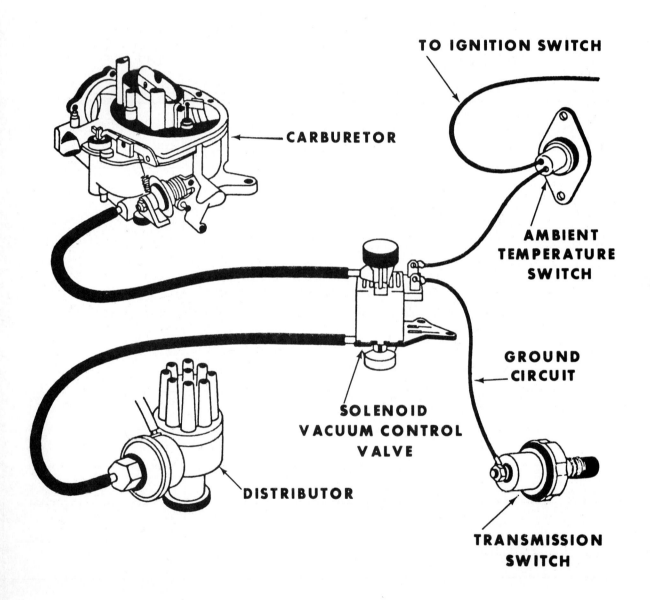

TRANSMISSION-REGULATED SPARK (TRS) SYSTEM

The transmission regulated spark system is another emission control system designed by Ford Motor Company to limit the formation of oxide of nitrogen exhaust emissions. The rather simple TRS system resembles, to some degree, Ford distributor modulator and electronic spark control systems, except that the electronic modulator and the speedometer-cable-driven speed sensor are not employed.

The TRS system is controlled entirely by transmission gear selection. The system employs a transmission switch that serves as a circuit grounding device by sensing either automatic transmission high-gear hydraulic pressure or manual transmission shaft linkage position, a solenoid-operated vacuum control valve, and an ambient air temperature switch. A spark delay valve may be used on some installations.

The system is designed to limit exhaust emissions by preventing vacuum spark advance in the lower gears. The fuel mixture and exhaust manifold temperatures are thereby increased by the retarded spark timing, effecting more complete combustion with a reduction in hydrocarbon and carbon monoxide emissions. When the transmission switch senses high-gear operation, the switch contacts are opened, deenergizing the vacuum control valve and opening the vacuum valve to allow manifold vacuum to be applied to the distributor advance unit. Vacuum spark advance is then applied in the usual manner.

If engine overheating occurs due to prolonged idling with retarded spark timing, the *ported vacuum switch* overrides the solenoid vacuum switch and applies vacuum to the distributor, regardless of transmission gear selection. The resulting idle speed increase assists in normalizing the coolant temperature.

At outside temperatures below approximately 49°F, the ambient air temperature switch contacts are open, deenergizing the TRS system and thereby overriding the solenoid vacuum switch. Vacuum is then applied to the distributor for vacuum spark advance regardless of transmission gear selection. The switch contacts close at temperatures above 65°F, energizing the system and returning control of the system to the transmission switch and solenoid vacuum switch. The temperature switch is positioned in a front-door pillar where it can sense outside air temperature without being influenced by passenger or engine compartment heat.

Testing of the transmission-regulated spark system should be performed by following the test sequence recommended by Ford Motor Company.

Figure No. 163

SPARK DELAY VALVE

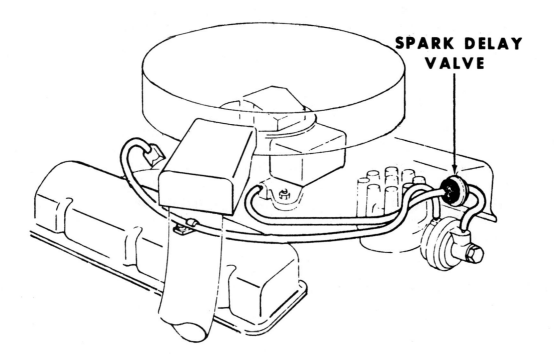

SPARK-DELAY VALVE

The spark-delay valve is another assist unit designed to control the ignition timing and limit the formation of exhaust emissions.

The valve is installed in the carburetor vacuum line at the distributor vacuum diaphragm on some Ford-built engines.

The function of the valve is to delay vacuum spark advance from occurring during rapid acceleration and to cut off the vacuum spark advance immediately on deceleration.

The delay valve cannot be tested or serviced. It must be replaced every 12,000 miles or 12 months, whichever occurs first. The length of the spark delay period varies with different engine applications. Several different valves are used and are color coded for identification. It is important that the replacement valve be the same color as the original valve.

When installing a new valve, the black-colored side of the valve must face the carburetor. The valve is designed for one-way operation and will not function if it is installed backward.

Figure No. 164

EXHAUST EMISSION CONTROL SYSTEM ASSIST UNITS

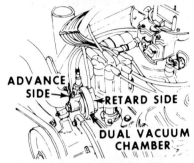

DUAL DIAPHRAGM SPARK CONTROL UNIT

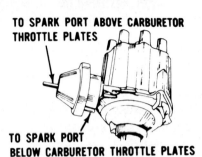

DOUBLE-ACTING SPARK CONTROL UNIT

THERMOSTATIC VACUUM SWITCH

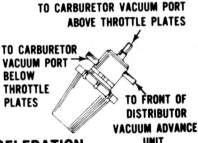

DECELERATION VACUUM ADVANCE VALVE

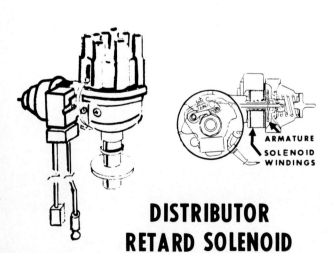

DISTRIBUTOR RETARD SOLENOID

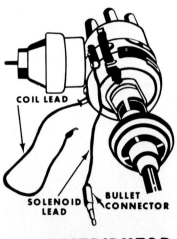

DISTRIBUTOR ADVANCE SOLENOID

© Copyright 1986, Tune-Up Manufacturers Institute

EXHAUST EMISSION CONTROL SYSTEM ASSIST UNITS

Dual-Diaphragm Vacuum Spark-Control Unit

The exhaust emission control-system on some engines requires a retarded spark timing at idle speed and during periods of deceleration. The spark retard permits the use of a slightly greater throttle opening at idle to allow for increased air intake. This action provides more complete combustion during the idle and deceleration periods of engine operation, which are the periods of greatest objectionable exhaust emissions.

The dual diaphragm distributor is used on some Ford Motor, General Motors, and American Motors engines. The diaphragm housing contains two spring-loaded independently operating diaphragms coupled by linkage to the movable distributor breaker plate. The forward diaphragm is moved by carburetor vacuum to shift the breaker plate against rotor rotation to advance the spark timing in the conventional manner. The rear diaphragm is actuated by intake manifold vacuum, which moves the breaker plate with the direction of rotor rotation to retard the spark timing.

The centrifugal (mechanical) advance mechanism is, of course, not influenced by the vacuum double-acting control.

The dual-acting control of the vacuum diaphragm is made possible by what is called a *ported vacuum advance* arrangement. A vacuum line is connected into the carburetor at a point above the throttle plate. Vacuum at this point, with a closed throttle, is very weak. During periods of idle and deceleration, intake manifold vacuum (which is high) is applied to the rear (retard) diaphragm, thereby providing the desired degree of spark timing retard. As soon as the throttle is opened, the high carburetor venturi vacuum is applied to the front (advance) diaphragm, and the spark timing is advanced in the conventional manner.

The dual diaphragm distributor actually affords three different phases of ignition timing. While the engine is being cranked for starting, timing may be for example a 6-degree before top dead center setting. As soon as the engine starts idling, the vacuum retard may set the timing at 6 degrees after top dead center. As the vehicle is accelerated, there will be varying degrees of spark timing advance, which will be a combination of the centrifugal and vacuum advance, depending on engine speed and load conditions.

Double-Acting Vacuum Spark Control Unit

Some Pontiac distributors used on exhaust emission controlled engines are equipped with a double-acting vacuum unit. This unit, however, contains only a single diaphragm; but in addition to the usual vacuum line fitting at the front of the unit, there is another fitting mounted on the rear of the unit. The vacuum line connections are the same as those used in the Ford dual diaphragm unit. With this arrangment the single diaphragm unit is used to retard or advance the spark timing by applying a vacuum source to either the rear or the front chamber of the vacuum unit.

© Copyright 1986, Tune-Up Manufacturers Institute

Thermostatic Vacuum Switch

As previously stated, many exhaust emission control system equipped engines idle with the ignition timing retarded. These engines also employ leaner calibrated carburetors. Because of the greater heat generated in an engine idling with retarded timing and lean fuel mixtures, there is a tendency for the engine to overheat during periods of prolonged idling, particularly during warm weather. To prevent this overheating possibility, a thermostatic vacuum switch is employed. This valve is also called a distributor vacuum control valve or a ported vacuum switch.

There are three vacuum ports on the switch. They are usually identified by letters such as D for distributor, C for carburetor, and M or MT for manifold (intake manifold). Vacuum hoses connect the switch ports to their respective sources. On some installations the lower port is fitted with a filter to sense ambient pressure.

The switch is mounted in the engine block water jacketing or in the water-distributing manifold where it can sense the temperature of the engine coolant.

When engine operating temperature is normal, the thermostatic switch does not function and does not influence ignition timing. But in the event the engine coolant temperature rises above normal (approximate 220°F), the switch automatically closes off the C port, shutting off the ported carburetor (retard) vacuum to the distributor, and opens the M port, applying full manifold (advance) vacuum to the distributor vacuum unit. The manifold vacuum applied to the distributor results in spark advance and an increase in engine idle rpm, with a corresponding increase in coolant circulation and fan action.

When the coolant temperature drops to normal value, the switch automatically shuts off the manifold vacuum to the distributor unit and reapplies the carburetor ported vacuum. The ignition timing now returns to its retarded setting, and the idle rpm drops to its slow idle speed.

Deceleration Vacuum Advance Valve

Another valve used in ignition spark control of some exhaust emission control equipped engines is a deceleration vacuum advance valve. Some engines may have a tendency toward a popping noise in the exhaust system during periods of deceleration "coast down" or during gear shifting. This condition is prompted by over retarding of the ignition timing.

To prevent this condition from occurring, the deceleration vacuum advance valve momentarily switches the vacuum to operate the distributor vacuum advance unit from its carburetor (low-vacuum) source to a manifold (high vacuum) for just a few seconds, and then back to its carburetor vacuum source.

The deceleration vacuum advance valve had popular application on 1966-1967 Chrysler-built CAP-equipped engines and a somewhat lesser application on 1968-1969 CAS-equipped engines. Some Pontiac and Ford models also employ this valve.

Distributor Retard Solenoid

An electric solenoid built into and controlling the action of the distributor vacuum unit is used on some 1970-1971 Chrysler-built V-8 engines. The function of the distributor solenoid is to provide spark timing retard during periods of idle and closed-throttle deceleration. This spark-retard action assists in the control of oxide of nitrogen exhaust emissions.

The distributor solenoid is energized and deenergized by the carburetor throttle stop (solenoid) unit and curb idle adjusting screw. The distributor solenoid is equipped with two electrical leads. The inner (feed) lead is connected to the field circuit of the alternator, and the outer lead is connected to ground at the carburetor throttle stop unit.

When the engine is idling, the curb idle adjusting screw contacts the carburetor throttle stop unit, completing the circuit that energizes the distributor solenoid windings. The magnetic field produced by the windings attracts the armature against the solenoid core. The armature, being connected to the vacuum diaphragm link, shifts the distributor breaker plate in the direction of cam rotation, thereby retarding the spark. As soon as the engine is accelerated, the idle adjusting screw breaks contact with the carburetor throttle stop unit, and the distributor solenoid circuit is deenergized. Vacuum spark advance then occurs in the usual manner.

On engine restart, either hot or cold, the retard solenoid will not be functioning since the throttle will be opened slightly to start the engine. The separated throttle contacts interrupt the solenoid ground circuit and deenergize the solenoid. Vacuum spark advance is then applied in the usual manner.

Distributor Advance Solenoid

Some 1972 Chrysler-built V-8 engines have a distributor equipped with a timing advance solenoid. The function of the solenoid is to provide a 7½-degree spark advance in the ignition timing during the engine cranking operation. This action promotes better engine starting.

The solenoid is internally positioned in the vacuum unit. The only visible part is a terminal connector on the side of the distributor housing. The short lead attached to the connector terminates in a male bullet connector, which also serves to identify the advance solenoid.

Power to energize the advance solenoid comes from the starting motor relay at the same connector that sends power to the starter solenoid. By using this type of connection, the distributor advance solenoid is activated only while the engine is being cranked. Current to the solenoid energizes a coil that actuates a link which moves the distributor breaker plate against cam rotation to introduce a 7½-degree spark advance. As soon as the engine starts and the ignition key is released, the ignition timing returns to the basic timing setting.

Figure No. 165

TESTING EXHAUST EMISSION CONTROL SYSTEM ASSIST UNITS

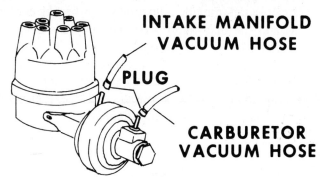

SET IGNITION TIMING
TEST CENTRIFUGAL ADVANCE MECHANISM

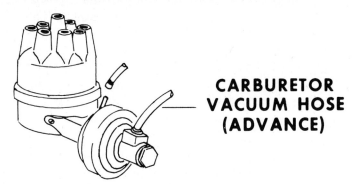

TEST VACUUM SPARK ADVANCE

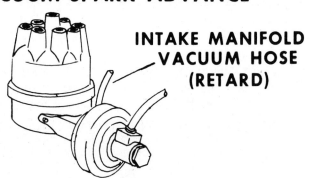

TEST VACUUM SPARK RETARD

© Copyright 1986, Tune-Up Manufacturers Institute

TESTING EXHAUST EMISSION CONTROL SYSTEM ASSIST UNITS

One of the most critical factors in engine tune-up is the testing of the distributor. If the centrifugal (mechanical) advance mechanism or the vacuum advance unit fail to function properly, the ignition timing will be late, acceleration will be poor, top speed performance will be unsatisfactory, and tendency toward overheating may occur. If either the mechanical or the vacuum advance systems should stick or freeze in the advance position, serious spark knock with all its engine-damaging effects will result.

Centrifugal Mechanism and Vacuum Unit Tests

1. With the engine at operating temperature, connect a power timing light and tachometer to the engine.
2. Disconnect and plug the distributor vacuum line. If the vacuum unit has dual diaphragm or has a double-acting diaphragm, disconnect and plug both vacuum lines.
3. Idle the engine at low idle speed. It may be necessary to reduce the rpm if the idle speed is over 600 rpm as a partial mechanical advance may be introduced at this speed.
4. Check the ignition timing setting by observing the timing marks. Reset the timing to specifications as required.
5. To test the centrifugal advance mechanism, accelerate the engine to 2000 rpm while observing the timing marks. Observe if a substantial amount of timing advance occurred when the engine was accelerated as indicated by the movement of the timing marks. The degrees of mechanical timing advance can be accurately tested if a delay-mechanism-equipped timing light is used.
6. To test the vacuum advance, lower the engine speed to 1500 rpm. While noting the position of the timing marks, connect the vacuum hose to the foward (advance) vacuum fitting. This will be the carburetor vacuum hose on dual-action vacuum units. As soon as the hose is connected, there should be an increase in the spark timing, as indicated by the timing marks.
7. To test the vacuum retard on dual-diaphragm or double-acting vacuum units, drop the engine speed to idle rpm. While observing the timing marks, connect the intake manifold vacuum hose to the rear (retard) vacuum fitting. The ignition timing should indicate a definite retard as soon as the hose is connected.

Trouble indicated in the centrifugal mechanism in step 5 requires disassembly and servicing of the distributor advance weights and cam mechanism. Lack of timing advance in step 6 or timing retard in step 7 are indications of defective vacuum diaphragms. Either condition requires the replacement of the vacuum advance unit.

Testing the Thermostatic Vacuum Switch

1. Connect a tachometer to the engine and bring the engine up to operating temperature. Observe the idle rpm.

2. Disconnect the intake manifold vacuum hose from the switch and plug or clamp the hose. There should be no change in the idle speed. If there is a drop in idle speed of 100 rpm or more, the switch is defective and should be replaced. Reconnect vacuum hose.

3. Next, cover the radiator sufficiently to induce a high-temperature engine coolant condition as indicated by the temperature gauge or warning light. *Do not* allow engine to overheat beyond this point. If the engine idle speed has increased by 100 rpm or more by this time, the switch is performing properly. If an increase in engine idle speed has not occurred, the switch is defective and should be replaced.

Testing the Deceleration Vacuum Advance Valve

The deceleration vacuum advance valve is tested by connecting a vacuum gauge to the distributor vacuum hose with a T that has the same inside diameter as the hose. Connect a tachometer to the engine, and tape the carburetor dashpot plunger so that it cannot contact the throttle lever.

1. Increase engine speed to 2000 rpm and hold the speed for approximately 5 seconds.

2. Release the throttle and observe the distributor vacuum. The vacuum should increase to more than 16 inches for a minimum of 1 second and fall below 6 inches within 3 seconds after the throttle is released.

If it takes less than 1 second or more than 3 seconds to obtain the correct vacuum reading, set the valve adjusting screw. Remove the plastic cover and turn the valve adjusting screw counterclockwise (in quarter-turn increments) to increase the time the distributor vacuum remains above 6 inches. Turning the screw clockwise will decrease the time limit. A valve that cannot be adjusted to specifications should be replaced.

Distributor Solenoid Tests

Distributor Retard Solenoid

The distributor retard solenoid can be tested for proper operation as follows.

1. Disconnect and plug distributor vacuum hose.

2. Start and idle engine. Check ignition timing to specifications with power timing light. Reset as required.

3. Disconnect retard solenoid connector at carburetor. *Note: Do not* attempt to disconnect lead at distributor solenoid.

4. Recheck ignition timing setting. Timing should advance from initial setting and engine speed should have increased. *Note:* If timing setting has not advanced, check for good ground contact at carburetor ground switch. If ground connection is good but timing does not advance after retest, retard solenoid is defective and must be replaced.

Note: Do not use jumper wires on solenoid or connect meters to the solenoid.

Distributor Advance Solenoid

Malfunction of the advance solenoid will result in hard starting. The action of the solenoid can be tested by using the following procedure.

1. Connect a tachometer to engine.
2. Disconnect and plug distributor vacuum hose.
3. Start and idle engine.
4. Disconnect solenoid lead bullet connector, which is above 6 inches from distributor. *Note:* Do *not* attempt to disconnect lead at distributor housing.
5. Clip a jumper wire to the solenoid lead male connector and make-and-break contact with other end of jumper lead to battery insulated post, while observing tachometer. Engine speed should increase about 50 rpm or more each time solenoid is energized. Replace a defective solenoid. *Note:* Avoid continuous application of battery voltage for periods exceeding 30 seconds to prevent possible damage to solenoid.

Note: If Chrysler electronic distributor advance solenoid is being tested, the single wire to distributor should be disconnected, *not* the double-bullet connector.

Deceleration (DECEL) Valve Tests

Malfunction Symptoms

Malfunction of the deceleration valve is usually indicated by either a rough idle or a high idle speed condition.

A rough idle can be the result of a lean fuel mixture caused by a leaking valve diaphragm, which permits the constant entry of additional air. If covering the small bleed hole in the bottom cover with a fingertip restores a smooth idle, a defective diaphragm is indicated and the valve should be replaced.

An excessively high idle speed of approximately 1200 to 1300 rpm can be caused by the valve being stuck open, permitting a constant draw of additional fuel from the deceleration section of the carburetor. This defect also requires valve replacement.

Test Procedure

The decel valve can be tested with the aid of a tachometer and a vacuum gauge.

1. Disconnect valve hose from carburetor; install a T fitting and connect vacuum gauge to T. Connect tachometer to engine.
2. Start and idle engine. Observe vacuum gauge reading, which should be zero vacuum. A vacuum reading indicates valve is stuck open.
3. Accelerate engine to 3000 rpm for a few seconds and release throttle quickly. Observe time required to drop vacuum reading to zero. Valve timing:
    1600-cc engine—3 to 5 seconds
    2000-cc engine—1 ½ to 5 seconds

© Copyright 1986, Tune-Up Manufacturers Institute

4. If time setting is out of limits, remove cap from top cover for access to adjusting screw. Slowly turn plastic adjuster counterclockwise a half-turn to lengthen time setting or a half-turn clockwise to shorten the time setting.

If proper time setting cannot be obtained, replace the valve.

Notes

Figure No. 166

CARBURETOR ADJUSTMENT

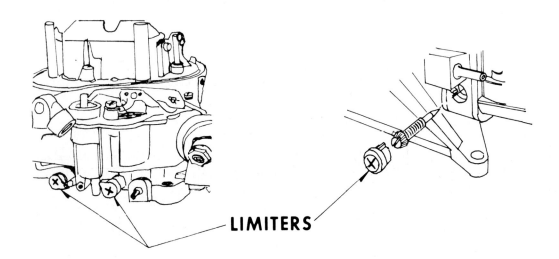

EXTERNAL TYPE LIMITERS

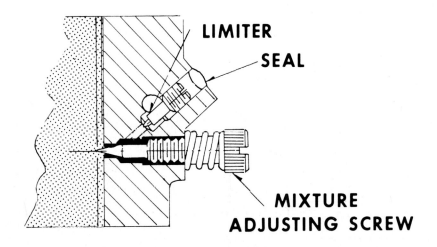

INTERNAL TYPE LIMITER

CARBURETOR ADJUSTMENT FOR EXHAUST EMISSION CONTROLLED ENGINES

As previously stated, the carburetors on emission controlled engines are calibrated for lean mixtures. By limiting the fuel mixture richness, the objectionable exhaust hydrocarbon emissions can be proportionately reduced.

Idle Limiters

To prevent the idle mixture adjusting screws from being accidentally or deliberately set too rich, many carburetors are presently being fitted with a device called an idle limiter. Idle mixture adjustment limiters are of two types, external and internal. That is, they are externally mounted on or internally positioned in the carburetor.

The external-type idle limiter is a plastic cap which fits over the head of the adjusting screw. The cap has a lug or stop projection which bumps the carburetor housing at each end of its travel, thereby limiting the amount of adjustment possible. Two-and four-barrel carburetors have a limiter on each mixture adjusting screw.

The internal-type limiter is located in a calibrated passage in the carburetor fuel idle circuit. The idle mixture is adjusted by turning the adjusting screw in the normal manner. But if the screw is backed out too far in an effort to enrich the mixture excessively, the overadjustment has no effect on the mixture setting after a certain point, since the richness of the mixture is then controlled by the size of the limiter passage.

It is very important that the idle limiters not be removed from the carburetor or mutilated to increase the adjustment range. If all the other engine components have been properly tuned prior to carburetor adjustment, a satisfactory mixture and idle adjustment can be effected with the range provided by the limiters. Since 1979, most carburetors have sealing caps over the idle limiter. These units are not adjusted in normal service.

Carburetor Lean-Roll Adjustment

The *lean-roll* method of setting the carburetor is presently a recommended method of adjustment. Basically, this adjustment is performed in the following manner:

1. Adjust the idle mixture screws, alternately, to obtain the highest idle rpm.
2. Adjust the idle rpm to the specified speed. If the carburetor is equipped with an idle stop solenoid, adjust the idle speed with the solenoid adjusting screw.
3. Turn one mixture screw clockwise slowly, and in small increments, until the idle speed drops about 15 rpm, or as recommended. Turn the other mixture to obtain another 15-rpm drop.
4. Readjust the slow idle speed to specifications.

If the carburetor is equipped with an idle stop solenoid, disconnect the solenoid lead and set the low (shutdown) idle speed with the carburetor idle adjustment screw. Reconnect the solenoid lead.

Note that the final mixture adjustment is performed by turning the mixture screw clockwise (leaner) to effect the lean-roll setting that permits the leanest air-fuel ratio consistent with effective control of emissions while providing acceptable engine performance.

Some carburetor mixture and idle adjustment specifications are listed as initial and final rpm settings. For example, turn mixture screws *in* until lightly seated, then back screws *out* four turns. Adjust carburetor idle speed screw (or idle solenoid screw) to obtain initial idle speed (say, 775 rpm). Then turn mixture adjusting screws *in* equally, in small increments, until final idle speed is obtained (say, 700 rpm). Electrically disconnected solenoid, if so equipped, and adjust carburetor idle speed screw to obtain shutdown idle rpm specified (say, 400 rpm). This new procedure sets the idle speed by adjusting the mixture screws. Following the car marker's recommended carburetor adjustment procedure "to the letter" is the best assurance that acceptable performance and effective emission control will both be achieved by the tune-up.

The correct carburetor adjustment procedure (along with the ignition timing and idle speed setting) is so important to effective emission control that starting with the 1968 models the car manufacturers have mounted a decal in each engine compartment listing these important specifications and outlining the carburetor adjustment procedure. Make it a practice to always follow these important recommendations.

It is also important that all special instructions be observed when adjusting the carburetor, such as headlight on high beam, air conditioning unit turned *on* or *off*, hot idle compensator valve held closed, air cleaner on or off, and/or automatic transmission in neutral or drive. Always an important consideration is that the engine must be up to normal operating temperature before carburetor adjustment is performed.

The use of a combustion analyzer to measure the carburetor air-fuel ratio is presently being recommended by car manufacturers. Be sure to follow the equipment manufacturer's instructions when using the analyzer to secure accurate readings. It is important the carburetor mixture screws be turned no more than 1/16 turn at a time with a 10-second wait between adjustments. It normally takes this long for the meter to sense the change in mixture setting.

The use of the *exhaust gas analyzer* as a testing device is increasing in popularity. The analyzer measures the concentrations of hydrocarbons (HC) and carbon monoxide (CO) emissions in the exhaust gases. The hydrocarbon reading is a measurement of the unburned fuel, usually in parts per million (ppm), in the exhaust gases. The carbon monoxide reading is a relative measurement of combustion efficiency.

Like other specialized test equipment, the exhaust gas analyzer assists the tune-up technician in performing professional service by directing him to the possible causes of trouble. High hydrocarbon (HC) readings indicate such conditions as misfiring or fouled spark plugs, overadvanced ignition timing, defective breaker points, vacuum leaks, or disconnected vacuum lines. High carbon monoxide (CO) readings indicate carburetor misadjustment, rich choke setting or defective choke action, high float level, restricted or plugged PCV valve, or a dirty carburetor air cleaner. And if both hydrocarbon and carbon monoxide readings are high, both the ignition and the carburetion systems need attention, which usually means a major tune-up should be performed. Used both before and after a tune-up, the exhaust gas analyzer will reveal the presence of trouble and verify its correction.

As exhaust emission standards become mandatory throughout the country, the exhaust gas analyzer will become an essential piece of test equipment.

It is important to have your instrument tested for accuracy and recalibrated as required at specified intervals to insure accurate readings.

The Vital Importance of Precision Tune-Up on Emission-Controlled Engines

The various systems of automotive emission control are not difficult to maintain. There are, however, certain critically essential requirements.

Major tune-up *must* be performed at the time or mileage interval recommended by the vehicle manufacturer. The engines must be *precisely* tuned to the car maker's specifications with the use of *quality* test equipment. Never before has *precision* tuning been such an absolute requisite.

Tune-up now serves two important functions: it must effectively control engine emissions, and it must keep the car owner pleased with overall engine performance. Only a professional precision tune-up can satisfy both requirements.

A very important point to remember when tuning emission-controlled engines is that there is no single unit or device that is the major controlling factor in tuning the engine. Every engine accessory or assist unit is designed to function in relation to other units. It is important that every step in the tune-up procedure be meticulously performed. There are no shortcuts, no minor tune-ups. Even minor deviations from some specifications can have undesirable effects on emission control.

An important fact for the tune-up technician to remember is that even after a tune-up the modern emission controlled engine may not have the performance capability and the smooth idle of former models. The reason is that air-fuel mixtures and ignition timing have been altered considerably to minimize air pollutants from the automobile engine. This minor inconvenience is very small payment for the cleaning of the air, and is an effort in which *all* motorists must participate as more states pass, or contemplate passing, clean-air legislation. It is particularly important that the carburetor be precision adjusted to limit exhaust emissions, rather than to effect a smooth idle.

© Copyright 1986, Tune-Up Manufacturers Institute

Notes

10

COMPUTERIZED ENGINE CONTROL SYSTEMS

Figure No. 167

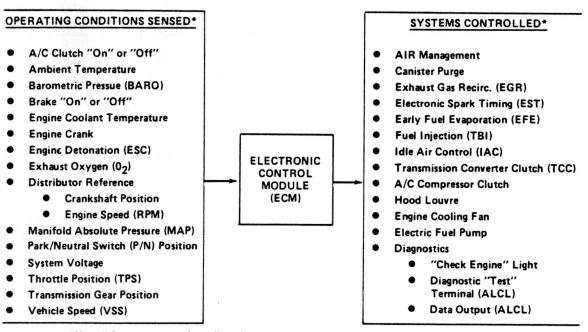

(Published by permission of General Motors Corporation.)

INTRODUCTION TO COMPUTERIZED ENGINE CONTROL SYSTEMS

To conform with the demands of strict emission control and fuel economy requirements, vehicle manufacturers had to improve the efficiency of their powerplants. Since high technology was readily available to meet this challenge, the computerized automobile made its debut in the late 1970s. The computers ability to improve every aspect of engine performance was so impressive that virtually every automobile built since 1980 has some type of microprocessor control governing ignition timing, fuel mixture, and emissions. The use of these control systems is growing with each new year and model introduction. The computer has already found its way into transmissions, suspensions, brakes, and comfort systems.

It would be foolish to deny the complexity of these systems. They are complex and will require strong technical background in the following areas:

1. Electrical and electronic circuits
2. Fuel systems
3. Ignition systems
4. Emission control systems

This chapter will deal with the operation of various systems, and present a simplified explanation of computer inputs and outputs. To fully understand the following information, a thorough understanding of the preceding chapters is necessary.

Figure No. 168

COMPUTER COMMAND CONTROL

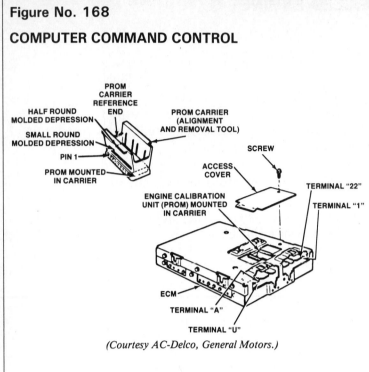

(Courtesy AC-Delco, General Motors.)

Figure No. 169

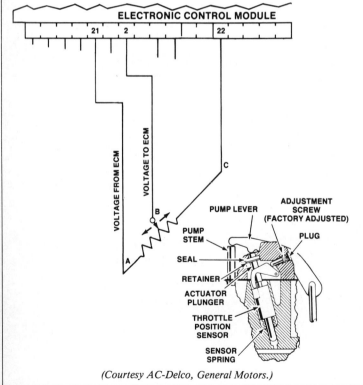

(Courtesy AC-Delco, General Motors.)

Figure No. 170

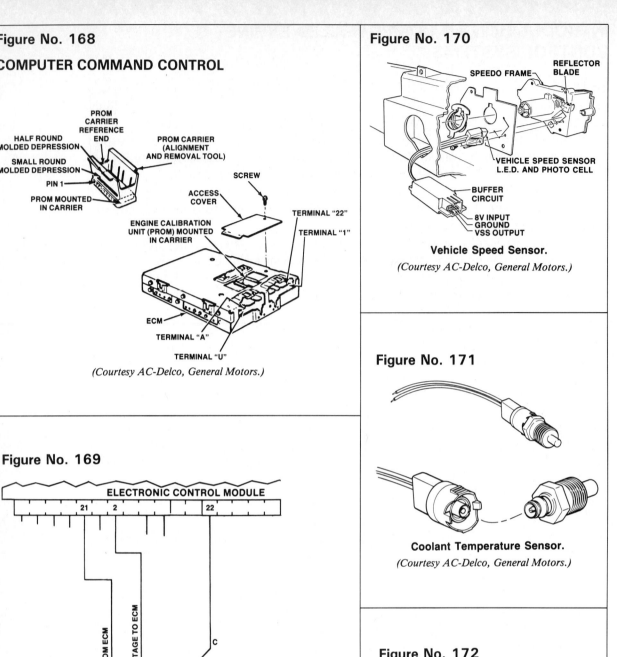

Vehicle Speed Sensor.
(Courtesy AC-Delco, General Motors.)

Figure No. 171

Coolant Temperature Sensor.
(Courtesy AC-Delco, General Motors.)

Figure No. 172

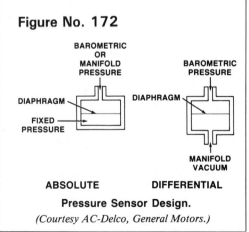

Pressure Sensor Design.
(Courtesy AC-Delco, General Motors.)

GENERAL MOTORS COMPUTER COMMAND CONTROL (CCC)

General Motors initiated computerized engine control applications during the 1980 model year. By 1982, virtually every GM passenger vehicle was equipped with one of three systems: full function CCC, minimum function CCC, or digital fuel injection (DFI) for the Cadillac Division. It is important to understand that any one system will vary greatly between year, make, and model application. These systems are in a constant state of gestation and are undergoing continuous changes in component configuration (hardware). Computer programming (software) is also updated periodically.

Electronic Control Module (ECM)

All computer systems consist of a series of inputs and outputs with the microprocessor at the center. General Motors calls the computer an electronic control module (ECM) and locates it within the passenger compartment. Although the ECM is constructed of many complex electronic components, it merely receives data from sensors, makes a decision, and issues a command in the form of an on-off signal. The ECM is a digital-type microprocessor, only capable of reading or issuing on-off signals.

The ECM memory is programmed with voltage and time values specifically tailored to particular vehicle applications. For example, if the computer receives a proper voltage for a correct length of time, it will issue a certain command. To make accurate decisions, the ECM must receive precise voltage signals from the engine sensors. To accomplish this, the ECM sends a constant 5-volt reference signal to the sensors. As the resistance values of the sensors change, the portion of the 5-volt signal that returns to the ECM is read as input data. The ECM maintains the constant 5-volt reference regardless of charging system fluctuations.

The ECM operates the computer command control (CCC) system in five modes:

1. *Shutdown mode:* When the engine speed drops below 200 rpm or charging system voltage is less than 9 volts, the ECM deactivates the mixture-control solenoid.
2. *Start-up Enrichment Mode:* After start-up, the ECM issues a rich command to the mixture-control solenoid. The length of this rich command is determined by data received from the coolant temperature sensor. The colder the engine, the longer the command. The start-up mode overrides all other modes.
3. *Open-loop mode:* During this mode, the ECM does not use the oxygen sensor for mixture control. Mixture is governed primarily by coolant temperature and manifold vacuum. The air switching solenoid is energized and air is injected upstream of the oxygen sensor. Cold engines operate in open loop.
4. *Closed-loop mode:* While in closed loop, the ECM is using the oxygen sensor to control the fuel mixture. To enter this mode, the following conditions must be satisfied:

    a. Hot oxygen sensor.
    b. Engine coolant temperature above calibration.
    c. Start-up enrichment time completed.

5. *WOT mode:* Wide-open-throttle mode overrides the open- and closed-loop modes. When the throttle is opened to greater than 86 percent, the ECM supplies a steady enrichment command to the mixture-control solenoid.

The ECM also contains circuits that enable the computer to perform self-diagnostics. When sensor circuits do not conform with the voltage and time parameters of the ECM program, the ECM recognizes a malfunction. Should a malfunction occur, the ECM turns on the dash-mounted CHECK ENGINE light and stores a trouble code in its memory. The CHECK ENGINE light remains on all the time the fault is present and is turned off if the problem goes away. If the problem is intermittent, causing the CHECK ENGINE light to flash on and off, and trouble code is stored in the memory for 50 starts of the engine. The intermittent trouble code is erased from the memory if the problem does not reoccur within the 50-start-time period. The ECM memory is also erased any time power to the computer is cut off. This can be done by pulling the ECM fuse or by disconnecting the battery. After servicing a CCC system, the mechanic should clear the memory.

Programmable Read Only Memory (PROM)

To tailor the computer to a specific vehicle application, a replaceable calibration computer chip is plugged into the ECM. This calibration chip is called the PROM. The PROM contains such information as inertial weight class, engine code, transmission type, accessories (A/C, P/S), final drive ratio, and tire size. Changing any of these component systems would alter the ECM calibration and affect performance and emissions.

A replacement ECM is not equipped with a new PROM. The PROM is a very delicate piece and must be handled with great care when being removed or installed. The ECM is programmed to identify a faulty PROM and will store a trouble code and turn on the CHECK ENGINE light in the event of a PROM failure.

Inputs

The ECM relies on sensor inputs to make command decisions. These inputs allow the ECM to control emission devices, timing advance, fuel mixture, idle speed, and transmission torque converter lock-up. When any sensor sends false messages, the computer can sometimes recognize the data as incorrect and set a trouble code, but other times it accepts the false sensor signal and issues commands that adversely affect engine performance. The most common causes of sensor trouble codes are non-computer-control components. For example, a defective purge control valve sets a rich oxygen sensor code. There is nothing wrong with the oxygen sensor, but a mechanic may assume that the sensor is at fault because the computer contains a code.

The *coolant temperature sensor* (Figure No. 171) is a thermistor threaded into the water jacket. As coolant temperature rises, the sensor resistance decreases. Sensor resistance increases as temperature falls. The ECM is programmed to read these resistance values as specific coolant temperatures. The ECM uses coolant temperature data to:

1. Vary fuel mixture while in open-loop and start-up enrichment modes.
2. Switch solenoids for control of air management, EGR, purge, and EFE.
3. Vary spark advance.
4. Vary idle speed.

Pressure sensors (Figure No. 172) supply manifold vacuum and barometric pressure information to the ECM. These sensors contain a flexible resistor that varies resistance with pressure changes. The ECM sends a fixed 5-volt reference value to the sensor, which returns to ground at the computer. The sensor signal wire is attached to the flexible resistor and transmits the resisted voltage signal to the computer. The ECM is programmed to read this return voltage as a specific pressure condition. There are two types of pressure sensors, absolute and differential.

The diaphragm of absolute pressure sensors is exposed to a fixed reference pressure on one side and to monitored pressure on the other. A manifold absolute pressure (MAP) sensor is connected directly to manifold pressure via a vacuum hose. Changes in manifold vacuum flex the diaphragm against the fixed reference pressure. As manifold vacuum rises, the voltage signal to the ECM decreases. The barometric pressure sensor (BARO) is identical to the MAP sensor, except it is exposed to atmospheric pressure instead of manifold pressure.

Differential pressure sensors compare manifold vacuum to atmospheric pressure. One side of the diaphragm is exposed to manifold vacuum, while the other side is open to atmosphere. Vehicles using this vacuum (VAC) sensor do not require a separate baro input because manifold vacuum signals are always being compared to ambient pressure conditions. The VAC voltage signal increases as manifold vacuum rises.

The ECM uses manifold pressure data to control fuel mixture and timing advance. Barometric pressure information is needed to modify the manifold vacuum signal in order to compensate for changes in altitude and weather conditions.

The *throttle position sensor* (TPS) (Figure No. 169) is a variable resistor attached to the throttle linkage. As the throttle valve opens and closes, the resistance value of the sensor changes. The change in resistance affects the 5-volt reference signal supplied by the ECM. At idle, the TPS signal is approximately 0.5-volt, and at wide-open throttle (WOT) it rises to about 4.5 volts. The computer relies on TPS data to control fuel mixture and the activation of various emission control solenoids. The TPS is the only sensor in the CCC system requiring adjustment. Should the TPS be out of range, hesitation would result.

The *vehicle speed sensor* (VSS) (Figure No. 170) is a fiber-optic device used to supply vehicle road speed (mph) information to the ECM. The sensor pick-up is located in the backside of the speedometer cluster. A light-emitting diode aims an invisible infrared beam at a rotating reflective surface in the speedometer housing. As the beam reflects off the rotating surface, it strikes a light-receiving diode positioned next to the LED. The blinking infrared signal is electrically processed into an on-off digital signal for the ECM. As road speed increases, the digital signal frequency increases, allowing the ECM to determine mph. VSS data are used to control torque converter lock-up.

Oxygen sensor information is used by the computer when operating in the closed-loop mode. A detailed description of oxygen sensor operation is found in Chapter 9.

Engine speed data (rpm) is received through the reference signal line from the distributor-mounted HEI module. A detailed explanation of the electronic spark timing (EST) system can be found in Chapter 7.

The CCC system may have several other inputs depending on how a particular vehicle is equipped. Cars with automatic transmission will have a *park/neutral switch* and a *top-gear switch*. The park/neutral switch is similar to a neutral safety switch and informs the computer when the transmission is in park or neutral. Gear position data are used by the ECM to affect idle speed and ignition timing. The top-gear switch is used to determine torque converter lock-up. If equipped with air conditioning, there will be sensor switches to indicate A/C clutch activation and high-side pressure problems. Four-cylinder cars with power steering will have a pressure switch in the power-steering hydraulic line. This switch allows the ECM to raise idle speed in compensation for excessive power-steering pump loads.

Outputs

Upon analyzing all the input data, the ECM makes decisions and issues commands in the form of on-off signals. To do this, the ECM acts like a giant ground base for various control solenoids. Twelve volts is sent to each control solenoid when the ignition is switched on. The ground for the solenoid windings is connected to the ECM. By completing or opening these individual ground circuits, the computer energizes or de-energizes the respective solenoid. The section of the ECM responsible for "making" and "breaking" the ground is called the *driver section*. The drivers are arranged in groups of four called *quad-drivers*. The drivers can be damaged by disconnecting or reconnecting the solenoid terminals with the key turned on. The drivers can also be destroyed if the windings resistance of a solenoid drops below 20 ohms. The driver section cannot be serviced separately, requiring ECM replacement should driver failure occur.

A control solenoid is generally used to control vacuum flow to a mechanical control device, such as an EGR valve. The mixture-control solenoid and fuel-injector solenoids control fuel directly. Some controls are electrical relays, which are actually a solenoid used to control a high-current-flow circuit. The number and types of output controls used in the CCC system will vary by vehicle model application and accessory equipment. The following is a brief description of those controls most commonly found.

1. The *mixture-control solenoid* is located in the carburetor and is used to control fuel flow and air bleeding to the main well. Refer to the carburetor section of this text for a detailed explanation.
2. *Purge-control solenoids* determine when the vapor recovery canister receives vacuum for purging into the intake manifold.
3. There may be one or two *EGR solenoids*. Energizing and de-energizing of the solenoid(s) control vacuum signals to the EGR valve.
4. *Air-management systems* usually have two solenoids: *air control* and *air switching*. The air-control solenoid directs air-pump supply either to the air cleaner or to the air-switching solenoid. The air-switching solenoid sends air upstream (exhaust header) or downstream (catalyst) of the oxygen sensor.

5. An *idle speed control (ISC) motor* is used to adjust idle speed. It is a dc motor that acts directly on the throttle linkage. By reversing polarity, the ECM extends or retracts the "nose" of ISC motor to maintain the factory-programmed idle speed regardless of engine idle load. It is important to note that the mechanic cannot adjust the idle speed. The idle speed is determined by the PROM. The ISC motor also contains an on-off switch. When the throttle is opened, the switch closes, telling the computer to fully extend the motor's "nose." Upon releasing the throttle, the linkage strikes the "nose," opening the switch. The ECM then slowly retracts the motor, seeking programmed idle speed. This operation acts as a dashpot function.

6. A *transmission converter clutch (TCC) solenoid* is located on the automatic transmission. When energized by the ECM, it connects the flywheel directly to the output shaft of the transmission through the torque converter. This acts as a computer-controlled overdrive by reducing slippage losses in the converter, which increases highway fuel economy. The ECM relies on information supplied by the vehicle speed sensor (VSS), coolant temperature sensor, throttle position sensor (TPS), and gear position switches for control of the TCC solenoid.

7. When the computer is ready to assume responsibility for ignition timing, it sends a 5-volt bypass signal to the B terminal of the distributor-mounted HEI module. This 5-volt output connects the ECM to the negative side of the ignition coil through the EST wire. Further details of the EST CCC system are found in Chapter 7.

8. Various control relays are used to control such items as electric early fuel evaporation (EFE), heating element grid, air-conditioning compressor clutch, electric engine cooling fans, and electric fuel pumps.

Figure No. 173

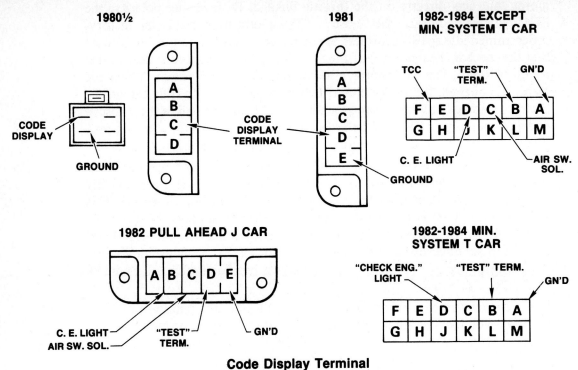

Code Display Terminal
(Courtesy AC-Delco, General Motors.)

Figure No. 174

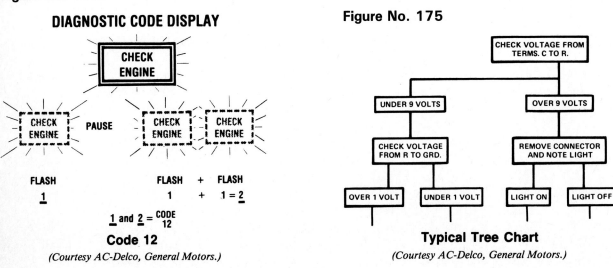

Code 12
(Courtesy AC-Delco, General Motors.)

Figure No. 175

Typical Tree Chart
(Courtesy AC-Delco, General Motors.)

GENERAL MOTORS CCC SYSTEM DIAGNOSTIC CODES

The electronic control module (ECM) is capable of recognizing faults in the sensory and mixture control circuits. Problems in the monitored circuits set a trouble code in the computer memory and turn on the CHECK ENGINE light. If the problem is intermittent, the CHECK ENGINE light goes out, but the trouble code is stored in the memory. The intermittent trouble code is erased from the memory if the fault does not reappear after 50 starts of the engine or if power to the ECM is disconnected. The proper method of erasing the memory is to remove the ECM fuse from the fuse block for 10 seconds or longer. It is important that the ignition be switched off when removing the fuse, or ECM damage could result. After repairing a CCC system problem, it is a good practice to clear the memory by pulling the fuse.

The secret to accessing the ECM memory lies in the proper use of the *assembly line communications link* (ALCL). The ALCL connector contains terminals used on the assembly line to assure that the engine is running properly before leaving the factory. This connector is located under the driver's side of the dashboard or on the side of the fuse block. Use a jumper wire to connect the test terminal to ground after the ignition has been switched on. This procedure causes the ECM to enter the diagnostic mode. Whenever the test terminal is ground with the key on and engine not running, the computer will:

1. Flash a code 12 and all other stored trouble codes.
2. Cycle the mixture-control solenoid at a fixed 30-degree dwell.
3. Energize all ECM-controlled solenoids.
4. Run the ISC motor in and out.

While in the diagnostic mode, trouble codes are flashed through the CHECK ENGINE light. The first code flashed is 12, which is indicated by one flash, followed by a brief pause, then two flashes in rapid succession (Figure No. 174). After code 12 has flashed three times, all the trouble codes stored in the ECM memory will be displayed, starting with the lowest numerical codes. Each trouble code is flashed three times. When all stored codes have been displayed, code 12 will flash again, indicating "end of message." Should there be no trouble codes stored, the CHECK ENGINE light will continually flash a code 12 until the jumper wire is removed.

Code 12 means that the ECM is not receiving a distributor reference signal from the R terminal of the HEI ignition module. Since the engine is not running, no reference signal is being generated at the pick-up coil. Therefore, code 12 is not indicating a fault, but is instead confirming that the diagnostic circuits of the ECM are intact and capable of identifying other system problems. Always look for trouble codes first when servicing a CCC-equipped vehicle. Use of the computer's memory is helpful in saving diagnostic time and finding intermittent problems.

If trouble codes are displayed while in the diagnostic mode, the use of a vehicle repair manual is required. The manual contains diagnostic "tree" diagrams for each possible trouble code (Figure No. 175). These diagrams lead the mechanic through a step-by-step procedure for solving system problems. Remember, components *not* connected to the CCC system can cause false codes to be set. Check all those items on the car that could have caused the same problem on a non-computer-controlled engine.

On General Motors vehicles equipped with CCC and electronic fuel injection

(EFI), the CHECK ENGINE light serves a slightly different function. Jumping the ALCL test terminal to ground with the key on and engine running will:

1. Indicate open-loop mode by flashing the CHECK ENGINE light rapidly (two to three times per second).
2. Indicate closed-loop mode by flashing the CHECK ENGINE light slowly (one time per second).
3. Indicate extremely rich mixture if light stays on.
4. Indicate extremely lean mixture if light stays off.

This operation is referred to as the *field service mode* and takes the place of the system performance check connector found on carbureted CCC systems.

Notes

Figure No. 176

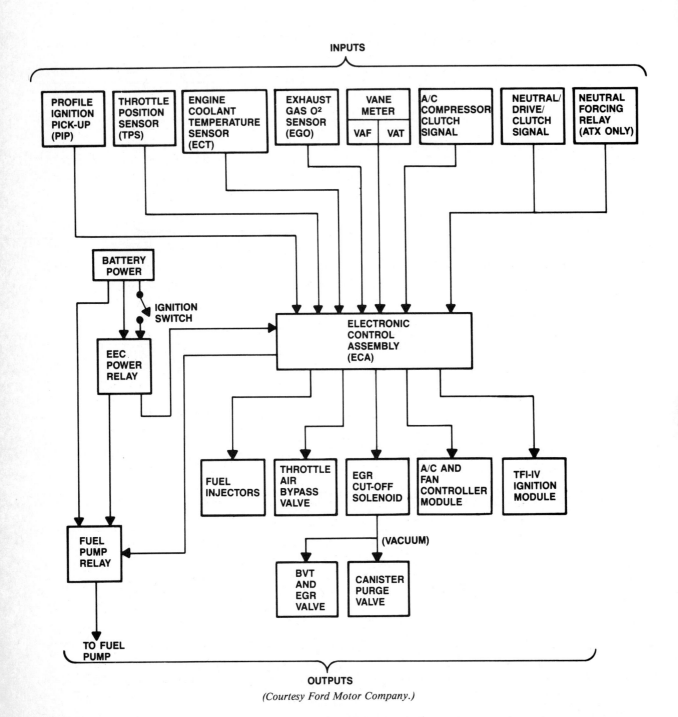

(Courtesy Ford Motor Company.)

FORD ELECTRONIC ENGINE CONTROL (EEC)

Ford introduced the electronic engine control (EEC) system in 1978 on the higher-priced Lincoln models and labeled it the EEC-I system. EEC-II was released in 1979 on various Ford and Mercury models. By 1980, EEC-III-controlled carburetor and fuel injected systems had replaced both EEC-I and EEC-II. In 1983, a Bosch multi-point electronic fuel injection (EFI) system was combined with electronic engine control to form the EFI-EEC-IV System. While all these numbers and abbreviations may seem complicated, the system is relatively simple and very similar to the General Motors CCC system. During the transition from EEC-I to EEC-IV, the basic configuration remained unchanged, with the addition of new inputs and outputs being the reason for nomenclature progression. Because the EFI-EEC-IV application is the most popular and contains the greatest number of inputs and outputs, it will be used to demonstrate EEC system operation.

The electronic control assembly is a microprocessor. As is the case with all computer systems, it is the control center for all operations. Ford calls its computer the electronic control assembly (ECA) and locates it within the passenger compartment. The ECA is constantly receiving input data from the sensor network, making decisions based on these data, and issuing commands to various engine controls. The ECA contains a calibration module, which adapts the control assembly to specific model applications and accessory options.

The programming of the ECA can best be described by dividing the functions into three separate operations: normal engine operation, cold or hot operation, and limited operation strategy (LOS).

The *normal engine operating* condition is classified by five modes:

1. The *crank mode* aids in engine starting by enriching the mixture and raising the idle speed. The EGR and canister purge solenoids remain closed, and timing is fixed at 10 degrees BTDC.

2. The *underspeed mode* is used to prevent stalling and help stabilize engine operation under adverse conditions. Fuel injector pulse width (opening time) is halved and idle speed is increased. The underspeed mode is entered anytime engine speed falls below 600 rpm.

3. *Closed throttle mode* covers idle and deceleration. Closed-loop fuel control is maintained at idle, and idle speed is held at the level programmed into the calibration module. Canister purge and EGR function are shut off. During deceleration, fuel supply is discontinued to reduce emissions and increase fuel economy. Fuel delivery is resumed just above idle speed or whenever the throttle is reopened.

4. The ECA enters the *part throttle mode* during cruise conditions with a warm engine. Fuel control is in closed loop. The idle speed control device is positioned to act as a dashpot if the throttle is closed. EGR and canister purge are functioning.

5. *The wide-open-throttle (WOT) mode* alters system outputs to give maximum engine performance. The ECA goes into open loop and enrichs the mixture. The EGR solenoid, air-conditioning clutch, and cooling fan are also turned off.

© Copyright 1986, Tune-Up Manufacturers Institute

Cold or hot engine operation changes the output signals to compensate for cold engine and overheat conditions. Depending on the severity of the temperature condition, the ECA will alter the following outputs:

1. Ignition timing advance schedule.
2. Fuel mixture strength.
3. Idle speed.

The *limited operation strategy (LOS)* is commonly referred to as the "limp-home" mode. An electronic failure in the system causes the ECA to fix all outputs at a limited performance level. This operation allows the vehicle to be driven at a reduced performance level until repairs can be facilitated. Fuel mixture is held constant, timing is fixed at 10 degrees BTDC, and emission control systems are shut down.

Inputs

The types of inputs used on the EEC system vary among year and model applications. When comparing the Ford and General Motors system, it may appear that GM has a more sophisticated and complex sensory network. Ford does not use a vehicle speed sensor, top-gear switch, or a power-steering switch. However, General Motors does not monitor intake air temperature or EGR valve position, but Ford does. General Motors has more accessory inputs, while Ford has more engine data inputs. Both systems are equally efficient and offer the same level of "high-tech" data processing.

The following offers a brief description of the sensor inputs that the ECA relies on to make accurate decisions:

1. *Crankshaft position* data are necessary for the determination of timing advance. Earlier EEC systems used a crankshaft position sensor located on either the harmonic balancer or at the flywheel flange. On these systems the distributor contained only secondary ignition components, and base ignition timing adjustment was fixed. EEC-IV systems use a Hall-effect distributor, which supplies the profile ignition pick-up (PIP) signal. This square-wave PIP signal supplies extremely accurate crankshaft position data to the ECA. Base timing is adjustable on EEC-IV. Further details on this system can be found in Chapter 7.

2. The *throttle position sensor (TPS)* is located on the throttle housing and connected directly to the throttle shaft. The TPS is a variable resistor. The ECA sends a 5-volt reference signal to the sensor and reads the resisted return voltage from the TPS signal wire. The return voltage ranges from 1.0 volt at idle to approximately 4.75 volts at wide-open throttle. The TPS is not adjustable.

3. The ECA receives data from two temperature sensors, the *engine coolant temperature (ECT) sensor* and the *air charge temperature (ACT) sensor*. On vehicles with multipoint fuel injection, the ACT sensor is called the *vane air temperature (VAT) sensor*. These inputs are thermistors. The ECA is programmed to read the returning voltage signals as specific coolant and intake

air temperatures. The ECT is a priority sensor used by the ECA to govern all outputs. The ACT and VAT are used to modify the spark-advance curve and to help determine intake air density for improved mixture control.

4. EEC applications with carburetors or throttle body fuel injection use *barometric and manifold absolute (B/MAP) sensors.* One housing contains the two sensors. They are both variable resistors receiving a 5-volt reference signal from the ECA. The barometric pressure sensor is open to atmosphere. The manifold vacuum sensor is connected directly to a manifold vacuum source. The information received from the sensors is processed by the ECA to control mixture, idle speed, timing advance, and various emission control solenoids. Later systems use one sensor for both functions. EFI-EEC-IV systems use a vane airflow (VAF) meter to measure intake air volume. Again, the computer sends a 5-volt reference signal to this potentiometer and reads the return voltage as an intake airflow volume. The greater the engine speed and throttle opening, the higher the intake air volume. This high volume of intake air causes the airflow sensor vane to move further open, sending a higher voltage signal to the ECA. Lower intake volumes send lower voltage signals. The intake air volume and temperature data are combined by the ECA to calculate mass airflow.

5. The *exhaust gas oxygen (EGO) sensor* is the same as those of other manufacturers. There is a detailed explanation of oxygen sensor operation in Chapter 9. On some V-8 models, two EGO sensors are used, one for each header bank. Ford also electrically heats the EGO on later applications. This is done to bring the vehicle into closed loop earlier in warm-up.

6. All EEC-I, II, and III systems used an *EGR valve position (EVP) sensor.* It is wired like a throttle position sensor and sends EGR opening information to the ECA. The ECA analyzes other inputs to determine what the EGR position should be.

7. When voltage is present at the air-conditioner compressor clutch, the same voltage is present at the *A/C clutch compressor (ACC) signal* wire to the ECA. Voltage in this wire causes the computer to raise idle speed in compensation for compressor load.

8. Gear position is indicated by the neutral start switch on automatic transaxles (ATX). Manual transaxles (MTX) use a clutch-engaged switch and transaxle neutral switch.

Outputs

The computer outputs are found in the form of solenoids and relays. With the ignition switched on, 12 volts is available to the solenoid windings. The ground for these windings is connected to the ECA. The ECA "makes or breaks" the solenoid grounds in order to control fuel mixture, idle speed, ignition timing, purge, and EGR.

Mixture is controlled by energizing and de-energizing solenoid type *fuel injectors.* The length of time the injector is held open (energized) determines the fuel flow to the engine. The length of opening time is called *injector pulse width* and is measured in milliseconds. The ECA controls pulse width by monitoring the following inputs:

manifold vacuum, barometric pressure, exhaust oxygen content, and coolant temperature.

Idle speed is maintained by the *throttle air bypass valve*. The bypass valve controls the flow of intake air through a passage that circumvents the throttle plate. The throttle air bypass valve is a solenoid device that varies the idle speed in relation to duty cycle signals received from the ECA. When energized, the solenoid totally opens the bypass passage, supplying maximum idle speed. When de-energized, the solenoid closes, allowing only the minimum throttle valve adjusted idle speed. The length of time the solenoid is energized, as opposed to the length of time it is de-energized, controls the idle speed.

A single solenoid is used to control vacuum signals to the EGR and canister purge valves. The *EGR SHUT-OFF SOLENOID* blocks vacuum when de-energized and opens when energized. Since the engine operating conditions required to permit EGR and purge operation are similar, the use of one solenoid instead of two is a logical simplification of the output network. Earlier EEC systems used much more complex controls, involving as many as three separate solenoids for EGR and purge. The solenoid remains off (closed) at idle, cold engine operation, and wide-open throttle. The solenoid is energized at off idle cruise on a warm engine.

Notes

Figure No. 177 Figure No. 178

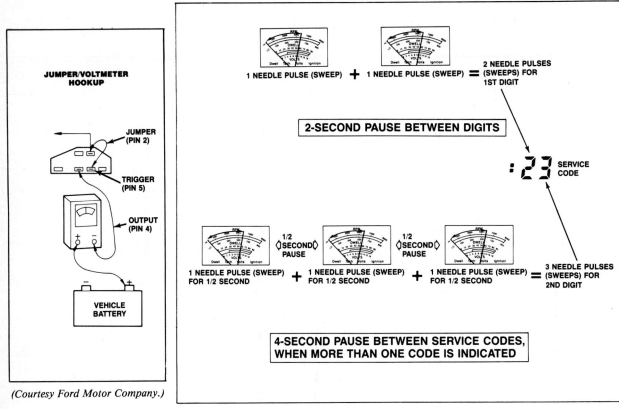

(Courtesy Ford Motor Company.)

(Courtesy Ford Motor Company.)

FORD EEC SYSTEM SELF-TEST CAPABILITIES

All Ford electronic engine control (EEC) systems contain a permanent memory that stores a self-test program. In the event of an electronic failure in the EEC system, the ECA enters the limited operation strategy (LOS) mode and greatly impairs the vehicle performance. The failure that causes the ECA to enter LOS can be determined by activating the self-test program and reading the digital output codes on an analog voltmeter.

A self-test connector is located in the engine compartment. Connect the voltmeter between the positive (+) terminal of the battery and the appropriate terminal of the self-test connector (Figure No. 177). Place a jumper wire across the test terminals of the connector as indicated by the repair manual. The ECA is ready to communicate faults that have been stored in the memory. The different self-test codes are displayed on the analog voltmeter in the form of needle sweeps (Figure No. 178). For example, the digit 2 is displayed by two sweeps of the voltmeter needle separated by a 1/2-second interval. After a 2-second pause, the digit 3 will appear in the form of three needle sweeps with 1/2-second intervals. Combining the digits, a code 23 is present, indicating that the throttle position (TPS) is out of range. All service codes are represented by a two-digit number. There will be a 4-second pause between service code displays.

Self-test operation is divided into three separate procedures. The different procedures allow the technician to better identify problem areas.

1. The *key on/engine off test* is performed by reading the voltmeter with the key in the run position and engine not running. This test compares sensor input values with calibrated sensor specifications stored in the ECA memory. Sensor inputs that do not agree with the ECA program set a code.

2. The *computed timing check* requires the use of a timing light. With the self-test jumper wire connected on a running engine, the ignition timing should be fixed at 30 degrees.

3. The *engine running test* checks both inputs and outputs under actual operating conditions. This test requires the technician to open and close the throttle in order to simulate changing conditions. The ECA is capable of comparing the different sensor inputs and determining if the proper outputs are attained.

Self-test hook-up and service codes vary among model and year of application. These tests require the use of a repair manual.

Figure No. 179

ELECTRONIC FEEDBACK CARBURETOR (EFC) SYSTEM

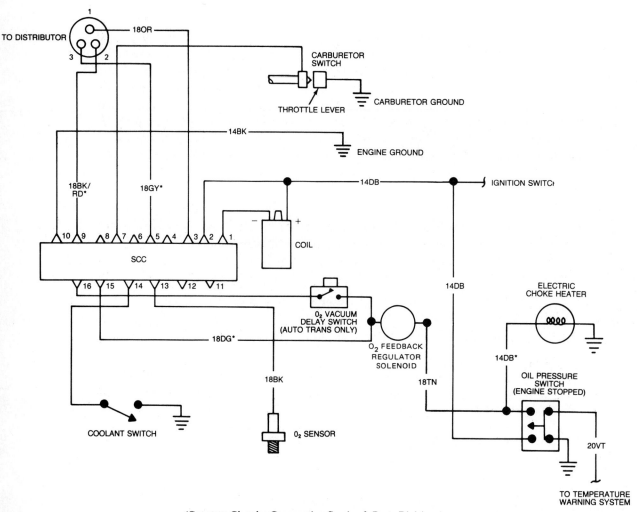

(Courtesy Chrysler Corporation Service & Parts Division.)

CHRYSLER ELECTRONIC FUEL CONTROL (EFC) SYSTEM

The Chrysler electronic fuel control (EFC) system was introduced on all 1981 model year passenger car applications. There may be visual similarities with earlier lean burn systems, but functionally the two systems are worlds apart. Lean burn systems affected only ignition timing and were based on an analog computer. EFC controls ignition timing, fuel mixture, and throttle position with a digital computer base. The spark control computer (SCC) is the command center for the system. The SCC issues commands based on information received from the sensor input network.

Inputs

A *Hall-effect pick-up* assembly located in the distributor supplies a digital square-wave signal to the computer. The computer uses this signal to determine engine speed (rpm), crankshaft position, and when the engine is being cranked by the starter. The SCC analyzes these data for control of spark advance.

A simple *coolant switch* tells the computer when the engine is hot or cold. When engine coolant temperature is below a calibrated value, the switch is closed. A closed switch causes the SCC to advance timing and enrich the mixture for improved cold engine drivability. After the engine has reached normal operating temperature, the switch opens. An open switch allows the computer to enter closed-loop operation and returns the timing schedule to a less advanced level.

A *vacuum transducer* is mounted on the computer and connected directly to manifold pressure through a vacuum hose. The transducer converts manifold pressure into an electrical signal. The computer is programmed to read the electrical signal as a specific manifold pressure valve. The transducer is used by the SCC to control both ignition advance and fuel mixture.

The *carburetor switch* is a crude type of throttle-position sensor. It is simply a throttle lever contact that completes a ground to the idle circuits of the computer when the throttle closes. A grounded carburetor switch causes the computer to cancel all spark advance and enter open-loop operation.

The *oxygen sensor* controls fuel mixture when in closed-loop operation. The sensor is located in the exhaust header and sends exhaust-gas oxygen content data to the computer.

Outputs

Computer output commands are divided into three separate categories: spark-advance control, fuel control, and throttle control.

The *spark control computer* maintains a preprogrammed amount of advance during cranking. Once the engine starts, spark advance is controlled by the computer based on inputs from the vacuum transducer and Hall-effect pick-up. As manifold vacuum rises, the timing advances, and if vacuum drops, timing is retarded. It should be noted that anytime the carburetor switch is grounded all spark advance is canceled for idle operation.

Fuel mixture is controlled by the computer through the O_2 *FEEDBACK SOLENOID*. The feedback solenoid is energized and de-energized by the SCC 10 times each second. Energizing the solenoid limits fuel flow and increases main well air bleeding, which yields a lean air-fuel mixture. De-energizing the solenoid restricts air bleeding and increases fuel flow, creating a rich condition. The length of time that the computer energizes and de-energizes the solenoid determines the mixture. This length of time is called the *duty cycle*. While in closed loop, the computer alters the duty cycle in relation to the signal received from the oxygen sensor. During open-loop operation, the duty cycle remains fixed.

The *throttle kicker solenoid* performs an electronic idle control function. The solenoid controls vacuum passage to a diaphragm-type throttle kicker. When energized by the computer, the solenoid sends vacuum to the kicker. The kicker extends against the throttle linkage, raising the idle speed approximately 400 rpm. The solenoid is energized when:

1. The air-conditioning compressor clutch is activated.
2. The heated rear window is turned on.
3. The throttle returns to idle. It will remain energized for 2 seconds, acting as a dashpot.
4. The engine is started and the coolant switch is open. It will remain energized for 30 seconds in order to clear fuel vapors generated by hot soak.

The EFC system does not contain any self-diagnostic capabilities. Its simplicity does not require the use of a memory or diagnostic harnesses. An electronic failure in the system causes the computer to enter a "limp mode." In "limp," the vehicle operates at greatly reduced performance levels, allowing the driver to bring the car in for repairs.

Notes

Notes

11

ELECTRONIC FUEL INJECTION

Notes

INTRODUCTION TO ELECTRONIC FUEL INJECTION

Prior to World War II, carburetion was the only form of fuel metering available for the gasoline-fired internal combustion engine. During the war, aircraft engines received mechanical fuel injection and the performance improvements were astonishing. Automotive engineers began applying this technology to the automobile by the 1950s. These early fuel-injection systems borrowed heavily from existing diesel technology and could only be found on exotic European sports cars. The diesel-type injection was extremely complex and cost far too much for widespread application. The 1968 Volkswagen Type III was the first production vehicle to be equipped with electronic fuel injection (EFI). This car is also noteworthy in that it was the first use of an on-board computer. EFI made it possible for all vehicles to gain the advantages of injection without the excessive cost and complexity.

A carburetor mixes fuel with air by discharging partially atomized gasoline into a low pressure zone of the intake air stream. The amount of fuel discharged is controlled by the mechanical action of throttle valves, accelerator pumps, chokes, and vacuum-controlled metering devices. Fuel injection discharges pressurized fuel and is controlled electronically. Throttle position and manifold vacuum are the primary metering inputs for a carburetor. EFI controls fuel flow by monitoring intake air volume or manifold vacuum, engine coolant temperature, intake air temperature, engine rpm, throttle position, and barometric pressure. All these variables affect the engine demand for fuel. Electronic carburetors also use these inputs, but lack the mechanical advantages of fuel injection.

Fuel-injection systems contain a pressurized fuel-delivery system. By opening the injector valves, the fuel is forced into the air charge and does not rely on venturi action to create a negative pressure zone. EFI does not require choke or accelerator circuits for additional fuel supply. Changes in altitude and weather condition do not affect engine performance. There is no need for intake preheaters or early fuel evaporation devices. The computer is able to compensate for changes in fuel demand by analyzing and reacting to the various sensory inputs.

Contrary to popular opinion, fuel injection is very simple to understand and service. The systems have been simplified over the years and do not require sophisticated equipment for diagnosis. Mechanics that are accustomed to working with fuel injection often complain that carburetor systems are more difficult to repair.

Figure No. 180

PORT INJECTOR SPRAY

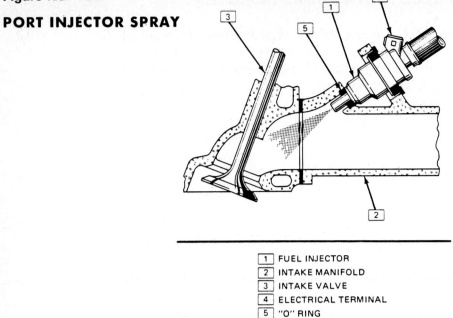

1. FUEL INJECTOR
2. INTAKE MANIFOLD
3. INTAKE VALVE
4. ELECTRICAL TERMINAL
5. "O" RING

(Published by permission of General Motors Corporation.)

Figure No. 181

TYPICAL FUEL DELIVERY SYSTEM

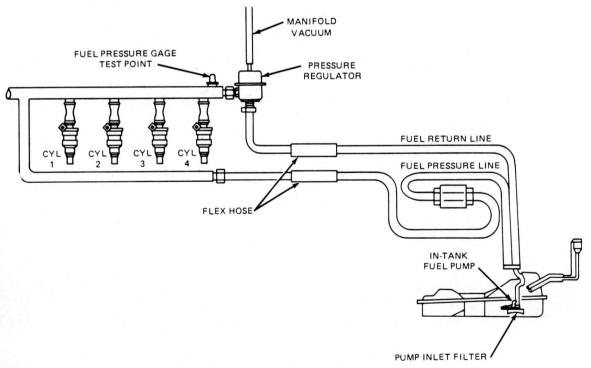

(Published by permission of General Motors Corporation.)

MULTIPOINT ELECTRONIC FUEL INJECTION

A multipoint fuel injection system (Figure No. 180) uses an injector for each cylinder. The injectors are located in the intake manifold and aimed directly at the backside of the intake valves. Multipoint injection is also called port or multiport fuel injection. Multipoint systems are considered the most efficient fuel systems available. These systems provide the following advantages:

1. Increased horsepower and torque output.
2. Improved fuel economy.
3. Greatly reduced emissions.
4. Smooth idle with flawless drivability.

Because of these advantages, every European import uses multipoint injection, and domestic manufacturers are rapidly converting to it.

All multipoint fuel-injection applications are derived from the Bosch Jetronic systems. Bosch manufactures the systems or sells the patent rights to vehicle manufacturers. The first EFI application was called the D-Jetronic or manifold pressure controlled (MPC) system. The D-Jetronic system has been replaced by the L-Jetronic (air flow control, AFC) system. The main difference in these systems lies in their sensor inputs. The MPC uses a manifold vacuum sensor, while the AFC uses an airflow meter.

Multipoint fuel injection is best understood by dividing it into three main service groups: electronic controls, fuel delivery, and air delivery.

Electronic Control Network

The computer is responsible for fuel control. Domestic cars use their existing computer systems for EFI control. For example, Ford connects the injection system to the electronic engine control (EEC) system, and General Motors incorporates the computer command control (CCC) system. Imported vehicles use a separate Bosch computer that is responsible only for fuel control. Regardless of the computer type, the engine sensor inputs provide information for precise fuel-injection control.

Manifold pressure data were used with early EFI systems. The sensor is connected directly to manifold pressure through a vacuum hose. As manifold vacuum changes, the resistance of the sensor return signal also changes. The computer is programmed to read the return signal as a specific manifold pressure. When vacuum rises, the amount of fuel supplied decreases. As manifold vacuum drops, the amount of fuel injected increases.

The manifold pressure sensor has been replaced by the *airflow meter*. Located upstream of the throttle valve, the airflow meter measures the volume of air drawn into the engine. The meter contains a measuring flap that swings open or closes in response to intake airflow. The flap is connected to a potentiometer (variable resistor), which varies the computer return signal in relation to the flap position. It is important to note that all intake air must pass through the airflow meter. The computer cannot supply fuel for air that enters the intake through vacuum leaks. For this reason, the

PCV system is sealed and does not have a breather element. Pull the oil dip-stick out while the engine is running and lean stumble will occur.

Intake air temperature is signaled by a variable resistor located in the intake air stream. As intake air temperature changes, the resistance of the sensor rises and falls. Intake air temperature data are used to determine intake density. The computer modifies manifold pressure or airflow readings in relation to intake air temperature values.

The *engine coolant temperature sensor* allows the computer to alter the fuel mixture for various temperature conditions. During cold engine operation, the mixture is enriched when the computer increases fuel supply.

The *throttle position sensor* is attached directly to the throttle shaft. The sensor contains two switch contacts and a variable resistor. The switch contacts are used to indicate closed throttle (idle) and full throttle (WOT). The variable resistor senses throttle opening changes. The throttle position sensor is adjustable and should be checked for proper operation as part of a normal tune-up.

Early EFI systems were equipped with rpm pick-up contacts in the distributor. This was a crude method of sending engine speed data to the computer for injector timing and selection. As the computers became more sophisticated, an rpm information lead was connected to the negative side of the ignition coil. The computer is capable of reading the primary circuit pulses as rpm in the same manner as an electronic tachometer.

The computer processes all the sensor input data and supplies the proper amount of fuel by controlling the electrical signal to the injectors. The fuel injectors are simple solenoids. When energized by the computer, the injector opens and sprays fuel into the port. The computer controls fuel supply by regulating how long the injector is energized (held open). The length of time that the injector remains open is called *injector duration* or *pulse width*. This time duration ranges from 2 to 5 milliseconds on most applications. A narrow pulse width yields a lean mixture. A wide pulse width creates a richer mixture.

The injectors are fired in pairs of complementary cylinders. For example, a four-cylinder engine with a 1-3-4-2 firing order has injectors grouped into a 1-4 and 3-2 opening sequence. This means that fuel is injected at approximately every top dead center (TDC).

During deceleration, the fuel injectors are held closed, shutting off the fuel supply. The computer identifies a deceleration condition when the:

1. Throttle position sensor indicates "closed throttle."
2. Engine speed is above the calibrated value.
3. Manifold vacuum or airflow is greater than the calibrated level.

Fuel injection is resumed when any one of these sensors indicates a need for power. Deceleration fuel shut off is responsible for reducing emissions and improving fuel economy.

Fuel-Delivery System

The fuel-delivery system (Figure No. 181) is responsible for maintaining a proper fuel pressure at the injectors. This pressure is called the *injector pressure*. A drop in fuel pressure leans the mixture. A rise in fuel pressure enrichs the mixture. The computer does *not* influence fuel pressure and controls injector duration on the assumption that the correct injector pressure is being maintained. Fuel pressure must be checked first whenever servicing any fuel-injection system.

Fuel-injected cars have at least one electric fuel pump and may have two pumps. Some applications have a chassis-mounted pump for pressure supply and a small in-tank unit to supply the chassis-mounted pump. A failure in either pump will cause a no-run condition. The fuel pump(s) is switched on by the fuel-pump relay, which is controlled by contacts in the computer, oil-pressure switch, and airflow meter. These switches open the fuel-pump circuit when the engine stops running, such as during stall conditions.

Pump pressure is fed into the primary delivery line, from which the injectors receive their fuel. Therefore, the pressure present in the primary delivery line is injection pressure. The fuel-pressure regulator is located at the end of the primary delivery line, downstream of the injectors. The regulator creates backpressure in the delivery line and is responsible for maintaining the proper injection pressure. The regulator consists of an enclosed diaphragm held in place by a calibrated spring. Pressure from the primary delivery line forces the diaphragm against the spring tension and opens a port to the fuel return line. Pressure in excess of calibration passes into the return line and flows back to the tank. Since the fuel pump exerts a pressure of 40 to 45 psi and the regulator maintains a backpressure of 30 to 35 psi, fuel is constantly flowing through the primary delivery and return lines.

Most pressure regulators are connected to manifold vacuum in order to modify the injector pressure in relation to changes in manifold vacuum. A vacuum-hose connection applies manifold pressure to the dry side of the regulator diaphragm. During high-vacuum conditions (idle), injector pressure declines. Low manifold vacuum (acceleration) increases injector pressure. On the average, primary fuel-delivery line pressure modulates between 25 and 35 psi.

Air-Delivery System

The air-delivery system is simply a manifold with a throttle valve located after the air filter. There are no preheaters or EFE devices, because only air flows through the manifold. Since fuel is delivered at the ports, there is no concern for fuel wetting in the manifold.

Idle speed is controlled by the air bypass circuit. Instead of changing throttle position to control idle speed, the throttle remains almost totally closed during idle. The bypass circuit consists of a passage that regulates the volume of air allowed to enter the intake. There is a bypass screw to restrict passage flow, which in turn adjusts idle speed. Do not move the throttle valve stop screw for idle adjustment or the throttle position sensor adjustment will be affected.

Figure No. 182

TBI UNIT

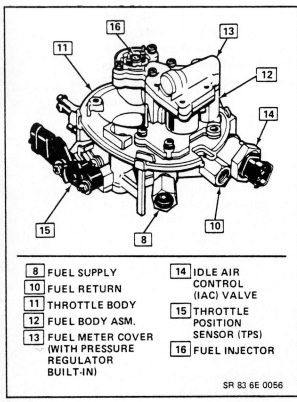

8	FUEL SUPPLY	14	IDLE AIR CONTROL (IAC) VALVE
10	FUEL RETURN		
11	THROTTLE BODY	15	THROTTLE POSITION SENSOR (TPS)
12	FUEL BODY ASM.		
13	FUEL METER COVER (WITH PRESSURE REGULATOR BUILT-IN)	16	FUEL INJECTOR

SR 83 6E 0056

(Published by permission of General Motors Corporation.)

Figure No. 183

TBI UNIT OPERATION

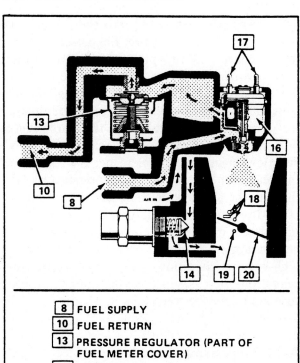

- 8 FUEL SUPPLY
- 10 FUEL RETURN
- 13 PRESSURE REGULATOR (PART OF FUEL METER COVER)
- 14 IDLE AIR CONTROL (IAC) VALVE (SHOWN OPEN)
- 16 FUEL INJECTOR
- 17 FUEL INJECTOR TERMINALS
- 18 PORTED VACUUM SOURCES*
- 19 MANIFOLD VACUUM SOURCE*
- 20 THROTTLE VALVE

*May Be Different on some Models.

SR 83 6E 0057

(Published by permission of General Motors Corporation.)

THROTTLE BODY FUEL INJECTION (TBI)

Throttle body fuel injection (TBI) is an interim stage between carburetion and multipoint fuel injection. While it does not have all the advantages of individual cylinder injection, it does a better job of metering and atomizing fuel than a carburetor. Throttle body injection allows vehicle manufacturers to make use of components and tooling used on carburetor applications. Most manufacturers are replacing TBI with multipoint EFI.

The TBI unit (Figure No. 182) contains one or two fuel injection solenoids. When energized, the injector solenoid(s) opens, spraying atomized fuel into the throttle body. The computer controls the volume of fuel by varying the injector pulse width. There is no choke valve. The computer senses engine coolant temperature and widens pulse width for rich mixture requirements. There is no accelerator pump. The computer reads throttle position and manifold pressure for load determination. Rapid throttle openings and sudden losses of manifold vacuum cause the computer to widen pulse width. During closed-loop operation, the oxygen sensor is monitored by the computer, which regulates pulse width in relation to exhaust oxygen content.

Idle speed cannot be adjusted. The computer is programmed with idle-speed parameters. There are two types of idle-speed control devices. Some systems use a dc motor, which can be extended or retracted by reversing motor polarity. The idle-speed control motor acts directly against the throttle linkage, opening and closing the throttle to maintain the programmed idle speed.

Other systems have an idle air bypass circuit cast into the TBI housing. These systems also have a computer-controlled motor. The motor extends or retracts a pintle that restricts the volume of air permitted to bypass the throttle valve.

The throttle body housing contains the fuel pressure regulator. The regulator is the same as the type found on the multipoint systems. A calibrated spring applies pressure to a diaphragm, which controls pressure bleeding into the fuel return line. It is not manifold vacuum controlled and maintains a constant fuel pressure at the injector.

Notes

12

TMI TUNE-UP PROCEDURE

Figure No. 184

TUNE-UP MANUFACTURERS INSTITUTE TUNE-UP PROCEDURE

1- **Cranking Voltage**
2- **Compression**
3- **Spark Plugs**
4- **Distributor**
5- **Dwell Angle**
6- **Ignition Timing**
7- **Carburetor**
8- **Charging Voltage**

© Copyright 1986, Tune-Up Manufacturers Institute

TMI TUNE-UP PROCEDURE

In Chapter 1, we discussed the five essential elements needed to perform quality tune-up. One of those essentials was a specific tune-up procedure.

Following a definite step-by-step procedure when tuning an engine and using the same procedure on every tune-up job is very important to consistently perform quality tune-up and effectively limit emissions.

Without a definite procedure, a tune-up man is apt to use a different approach or procedure on every engine. This leads to critical services being overlooked from time to time. It also leads to imperfect results after the tune-up is completed, necessitating rework and wasted time.

Besides, if the tune-up specialist uses the same procedure repeatedly, the operation becomes "second nature" and the jobs proceed smoothly and quickly. The trouble is found the first time around, which is the tune-up man's major objective. Further, the car owner can be assured his vehicle can pass a state emission test, if he is subject to a state inspection.

The Tune-up Manufacturers Institute Tune-Up Procedure contains the steps which are essential to a good tune-up in the sequence in which they should be performed.

1. Cranking voltage
2. Compression
3. Spark plugs
4. Distributor
5. Dwell angle
6. Ignition timing
7. Carburetor
8. Charging voltage

Figure No. 185

CRANKING VOLTAGE TEST

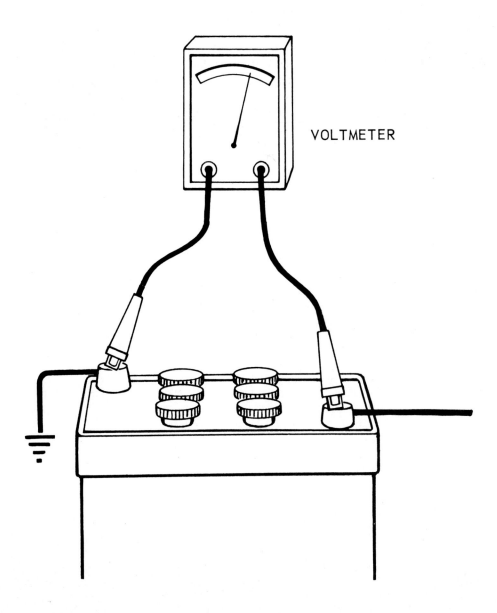

CRANKING VOLTAGE TEST

The cranking voltage test is one of the area tests used in this procedure. Area tests are used to quickly determine the condition of complete circuits. If the circuit being tested is within specified limits, it is logical to assume the individual components in the circuit must also be within limits and no further testing is necessary. If a variation from specified limits exists in a circuit test, the trouble must be located by testing each individual componenet in the circuit. Area testing is therefore an excellent, time-saving test procedure.

A cranking voltage test quickly determines whether or not sufficient voltage is being applied to the ignition system while the engine is being cranked. Should battery voltage drop below minimum standards while the engine is being cranked, the ignition system is being starved and the engine will not start. This is a very common cause of cold-weather starting difficulties.

To conduct the cranking voltage test, the voltmeter is connected across the battery. A jumper lead is used to ground the ignition primary circuit to prevent the engine from starting. With the ignition switch turned on and the engine cranking, observe the voltmeter reading and listen to the speed of the starting motor or observe cranking rpm on a tachometer. If the voltage is less than 9.0 volts for a 12-volt battery, or less than 4.5 volts for a 6-volt battery, trouble is indicated.

The cause of trouble may be a discharged or defective battery, defective battery cables, high resistance in battery cable connections, starting motor malfunction, or a defective starting motor switch or solenoid.

Figure No. 186

CHARGING VOLTAGE TEST

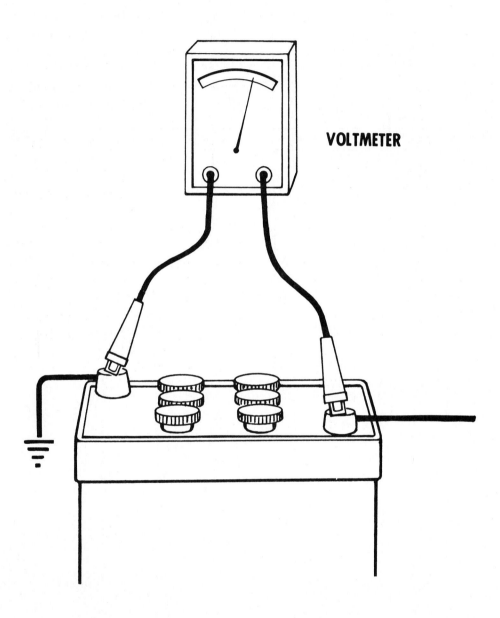

CHARGING VOLTAGE TEST

The charging system voltage test is another of the area tests. This test is a good indication of the overall operation of the electrical system, including generator output and voltage regulator setting. The test also indicates the voltage applied to the ignition system, which is an important factor in cases of breaker-point burning and when short operating life of other electrical units is experienced.

To conduct the charging voltage test, the voltmeter is connected across the battery. Operate the engine at a fixed speed of 1500 to 1800 rpm. Observe the voltmeter for the charging voltage reading. If there is a tendency for the voltage to climb slightly, wait for the highest reading. Compare the highest voltage reading to specifications.

The charging voltage specification for a 12-volt system is generally between 13.8 and 15.0 volts. The specification for a 6-volt system is generally between 6.8 and 7.5 volts.

Voltage readings higher than specified indicate a defective or misadjusted voltage regulator, high resistance in the regulator ground circuit, or a defective generator field circuit.

Voltage readings lower than specified indicate a loose drive belt, a defective generator, a defective or misadjusted voltage regulator, high resistance in the charging circuit, or a discharged battery.

It is assumed that a partially charged battery, which would cause a low false charging voltage test reading, would have been revealed in the cranking voltage test, the first test in the procedure. However, if it is suspected that an undercharged battery is causing the low-voltage reading, the battery can be quickly checked as being the cause of the trouble by the following method. Connect a test ammeter at the battery terminal of the regulator. If, during the voltage test, more than 10 amperes is flowing the battery is too low to make an accurate voltage test. To proceed in properly conducting the charging voltage test, either place the car battery on fast charge, replace the car battery with a fully charged battery to complete the test, or connect a ¼-ohm resistor in series with the battery to simulate a fully charged battery condition.

Voltage readings either below or above the specified setting indicate the need for a complete charging system test.

Figure No. 187

POWER LOSS

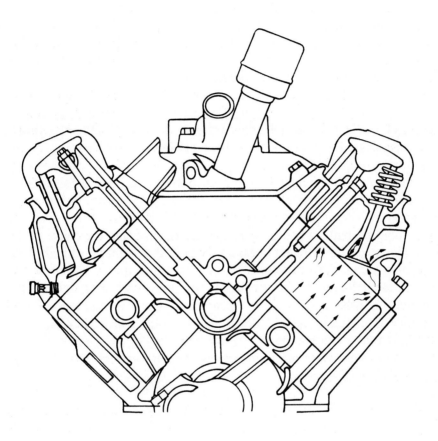

COMPRESSION STROKE LEAKAGE

POWER LOSS

Before a successful tune-up can be performed, it must be determined if the engine is in a satisfactory mechanical condition. An engine that has burned or leaking valves, worn piston rings, sticking valve lifters, leaking cylinder head gaskets, or other mechanical malfunctions will not perform efficiently even after being tuned-up. These conditions *must* be corrected before an engine can be tuned to perform satisfactorily or before emissions can be effectively limited.

Since the power developed by an engine on its power stroke is largely dependent on the efficiency of the compression stroke, and because of the testing convenience afforded, the compression stroke is used for testing engine condition.

During the compression stroke, the air-fuel mixture is compressed in the tightly sealed combustion chamber. Should any openings be created by burned valves, leaking gaskets, or worn piston rings, the reduced amount of the air-fuel mixture would proportionately reduce the power output of the engine.

Leakage at any point in the combustion chamber will affect efficient engine operation. Leaking intake valves will allow a portion of the air-fuel mixture to be pushed back into the intake manifold during the compression stroke, and less fuel will be available for the power stroke. During the power stroke, the expanding gases will leak past the burned valves, and less pressure will be available on the head of the piston. Also, burned gases will be forced into the intake manifold to mix with the air-fuel mixture. A diluted air-fuel mixture will then be available for the next intake stroke, and consequently less power will be developed by the engine. If the exhaust valve is burned, the expanding gases will leak through it, and less power will be available from the cylinder.

Any leakage past the piston rings will also affect the power of the engine. During the compression stroke, part of the air-fuel mixture will be forced into the crankcase and cause oil contamination. The power stroke will also force burned gases into the crankcase. These gases will overheat some of the oil, turning it to carbon, and the oil will become contaminated.

A leaking head gasket will permit water to be drawn into the cylinder during the intake stroke. During the compression and power strokes, gases will be forced from the combustion chamber into the cooling system and cause the engine to overheat. Also, a less dense air-fuel mixture will be available for the power stroke.

It is obvious that conditions of compression stroke leakage are proportionately reflected in engine power loss and must be corrected before an engine can be properly tuned.

Figure No. 188

COMPRESSION TEST

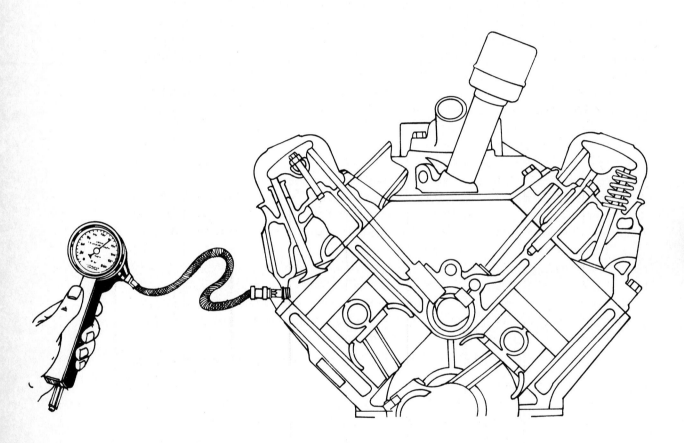

COMPRESSION TEST

A compression test is conducted to determine if the piston can compress the air-fuel mixture in the cylinder to a specified pressure, which is essential for efficient combustion and maximum power output.

The compression test results are compared to the specifications. The compression pressure specifications will be listed, for example, as 160 pounds per square inch (psi), plus or minus 10 percent:

$$160 \text{ psi} \pm 10\%$$

The plus or minus tolerance is an important part of the specification, since it limits the permissible variation between cylinder pressures. Only when *all* the cylinder pressure readings are within the limits of the specifications is smooth engine operation possible.

The figures in Figure No. 188 indicate an abnormally low reading in cylinder 1. A 10 percent reduction in the reading of cylinder 2 (160 − 16 − 144) reveals that there is more than the permissible pressure variation between cylinders 1 and 2. The cause of the low pressure in cylinder 1 must be corrected before an effective tune-up can be performed.

The compression test reveals the mechanical condition of the engine. Any condition that permits a loss of compression pressure has a decided and undesirable influence on the engine's power output.

When conducting a compression test, you are concerned with two values: the compression pressure in each cylinder as compared to specifications, and the permissible variation of pressure existing between cylinders.

Unless the individual cylinder pressures are up to specified limits, the engine cannot be properly tuned to maximum efficiency and performance. If a greater than permissible variation of pressure exists between cylinders, a degree of engine roughness will exist that your tune-up cannot correct.

The compression test is conducted in the following manner with the engine at normal operating temperature:

1. Carefully disconnect the spark-plug cables from the plugs.
2. Remove all the spark plugs.
3. Block the carburetor linkage to hold the throttle and choke valves wide open.
4. Connect remote-control starting switch. Observe precautions on next page.
5. Place the compression tester in cylinder 1 and crank the engine about four compression strokes; observe the first and fourth readings.
6. Crank each cylinder the same number of revolutions and record the readings.
7. Compare maximum readings and cylinder variations with specifications. Pay particular attention to the degree of variation between cylinders to make sure that a variation greater than permitted by specification does not exist. The figure in Figure No. 188 indicates an abnormally low reading in cylinder 1, since there is a greater than 10 percent variation between it and cylinder 2.

© Copyright 1986, Tune-Up Manufacturers Institute

If the compression builds up quickly and evenly to the specified pressure on each cylinder and does not vary more than the allowable tolerance, the readings are normal. The engine can be considered acceptable for tune-up.

Worn piston rings will be indicated by low compression on the first stroke, which tends to gradually build up on the following strokes. A further indication is an improvement of the cylinder reading when about a tablespoon of motor oil is added to the cylinder through the spark-plug hole with an oil can.

Valve trouble is indicated by a low-compression reading on the first stroke and pressure does not rapidly build up with succeeding strokes. The addition of oil will not materially affect the readings obtained.

Leaky head gaskets on two adjacent cylinders will produce the same test results as valve trouble. An additional indication of this particular trouble is the appearance of water in the crankcase.

Carbon deposits result in compression pressures being considerably higher than specified. It is possible that carbon can hide a defect in the cylinder, as the deposit will raise the compression pressure of a cylinder to the extent which might compensate for leakage.

Underhood Remote Control Starting Precaution

Starting in 1963, the ignition system circuitry of General Motors cars has been revised to bypass the ignition resistor during the cranking operation through a contact in the starter, instead of through the ignition switch as was used formerly. The change was made to overcome the possibility of a sticking switch condition which would cause distributor point failure.

The following precaution *must* be observed when connecting a remote control starting switch to a car employing the redesigned switch. Disconnect the distributor primary lead from the ignition coil and turn the ignition switch to the *on* position while remotely cranking the engine.

Failure to observe this precaution will result in burning out the ground circuit in the ignition switch. Follow manufacturer's specifications on cars equipped with solid-state ignition or with catalytic converters.

Notes

Figure No. 189

SPARK PLUGS

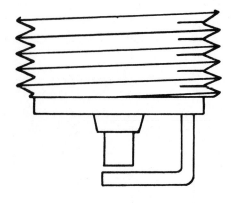

NEW PLUG ELECTRODES

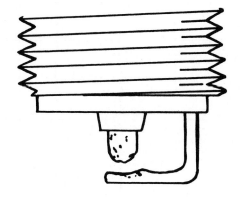

NORMAL ELECTRODE WEAR

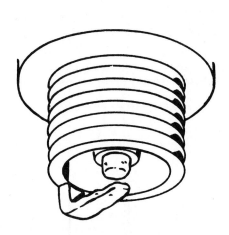

WORN OUT ELECTRODES

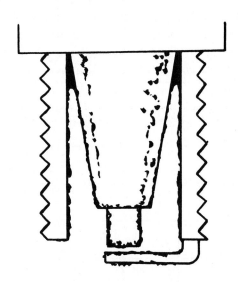

LEAD OR CARBON FOULING

© Copyright 1986, Tune-Up Manufacturers Institute

SPARK PLUGS

At the completion of the compression test, the spark plugs should be serviced next. A spark plug must operate in combustion temperatures as high as 4000°F at 1000 to 2000 sparks per minute, and withstand pressures as high as 800 pounds per square inch. After several thousand miles of service, the spark plug gap widens, and the electrodes become rounded due to the combined action of intense heat, pressure, corrosive gases within the combustion chamber, and spark erosion. The plug insulator also becomes covered with carbon and lead deposits to a greater or lesser degree.

The correct spacing of the spark plug gap is very important since it influences the entire range of engine performance: starting, idling, accelerating, and cruising. Further, the gap is also instrumental in controlling exhaust emissions. Uniformity of all spark plug gaps is extremely important for smooth engine operation. Always check the gap of new plugs against engine specifications before installing them.

To maintain efficient spark plug performance, plugs may be washed, cleaned, filed, inspected, and regapped at regular intervals. The electrodes should be filed flat because current is emitted from a sharp corner much more easily than from a blunt surface. If this maintenance is neglected, up to 30 percent more voltage is required to fire the spark plugs.

Due to the extreme importance of totally efficient ignition system performance in the effective limiting of exhaust emissions, many car manufacturers are recommending the replacement of spark plugs every 6000 miles if leaded gasoline is used and every 10,000 miles if unleaded (lead-free) gasoline is used. Follow manufacturer's recommended interval. Recent recommendations have been greatly increased. For example, some vehicles now recommend a change at each 22,000 miles.

If any question exists in your mind about the condition of the spark plugs you take from an engine, suggest to your customer that they be replaced with new plugs. The quick starting and added acceleration the car owner receives from new plug performance and the savings he will realize in greater fuel economy will shortly repay the cost of the new plugs. Needless to say, he will also be aiding in the fight for "cleaner air," since exhaust emissions are tremendously accelerated by misfiring spark plugs.

Use your specifications, and a cross-reference table if necessary, for proper sparkplug selection. Torque the plugs to specifications to insure proper heat transfer.

Figure No. 190

IGNITION TIMING

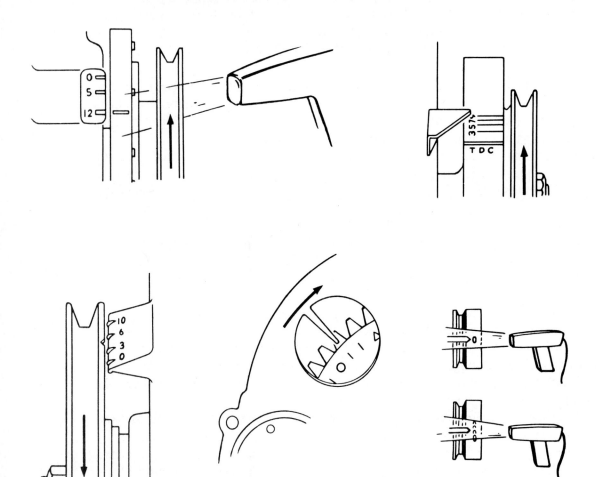
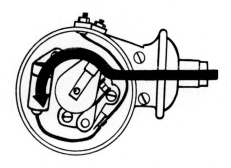
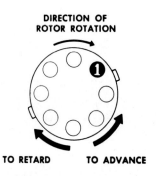

© Copyright 1986, Tune-Up Manufacturers Institute

IGNITION TIMING

Correct ignition timing is one of the most important factors relative to efficient and economical engine operation. If the initial timing setting is not correct, the entire range of the spark advance curve will be out of limits.

Ignition timing is checked and set with the aid of a power timing light. The light is energized by battery current and triggered by the voltage of the spark plug to which it is connected, usually the plug in cylinder 1.

The procedure for ignition timing an engine is as follows:

1. Locate the timing marks on the crankshaft pulley, harmonic damper, or flywheel. If they are not readily visible, wipe them with a cloth and mark them with chalk or paint.
2. Operate the engine until normal temperature is reached.
3. Stop the engine and connect a tachometer and a timing light. Disconnect and tape the distributor vacuum hose. If there are two hoses, disconnect and tape both hoses.
4. Start and idle the engine. The light will flash each time spark plug 1 fires.
5. Operate the engine at the specified speed and aim the timing light at the timing marks. *Caution:* Be very careful of the revolving fan blades.
6. Reset ignition timing if timing mark appears on either side of reference pointer. Reconnect distributor vacuum hose(s).

The ignition timing is adjusted by slightly loosening the distributor hold-down screw and slowly turning the distributor body against rotor rotation to advance the timing or with rotor rotation to retard the timing. When the specified mark is aligned with the pointer, securely tighten the distributor hold-down screw. Then recheck the alignment of the timing marks.

Figure No. 190 illustrates a method of easily determining the direction of rotor rotation without removing the distributor cap or cranking the engine by merely observing the postition of the vacuum advance unit on the distributor body. Rotor rotation can be determined by visualizing an arrow passing through the vacuum line and around inside the distributor cap. As illustrated, the rotor turns counterclockwise. Had the vacuum advance unit been positioned below the centerline of the distributor, rotor rotation would have been clockwise.

The timing mark should appear steady as the light flashes. If the mark appears to "fan out" or "wander" as the light flashes, trouble is indicated. This condition can be caused by pitted breaker points, misaligned points, improper point spring tension, loose or worn breaker plate, worn distributor shaft, worn distributor shaft bushings, or excessive lash anywhere in the distributor drive mechanism. These conditions must be corrected before an engine can be properly tuned.

After the initial timing is set, slowly accelerate the engine to approximately 1500 rpm while observing the timing mark with the light. The timing mark should move steadily away from the pointer. This is an indication that the spark advance mechanism is in operation. If there is little or no indication of spark advance, the distributor should be removed from the engine for a complete test.

© Copyright 1986, Tune-Up Manufacturers Institute

Connecting the Timing Light

The power timing light may be designed with a pick-up unit which is placed around the spark-plug cable, making an easy connection. In the event a timing light is used that must be clipped to spark plug 1 and the plug is not readily accessible, tower 1 in the distributor cap may be used with the aid of an adapter. The spark plug in the companion cylinder to No. 1 may also be used.

The companion cylinder to No. 1 can be quickly determined by the engine's firing order:

1. In a four-cylinder engine, the companion cylinder is the third cylinder in the firing order.
2. In a six-cylinder engine, the companion cylinder is the fourth cylinder in the firing order.
3. In an eight-cylinder engine, the companion cylinder is the fifth cylinder in the firing order.

The distributor cap tower of the companion cylinder may also be used if desired.

Any engine operating on the four-stroke cycle principle can be ignition timed from four locations.

Never puncture the spark-plug cables to hook up a timing light or other test equipment. Use the proper adapter.

Never power time an engine. Always use a timing light.

Ignition Timing of Exhaust Emission Control System Equipped Engines

Setting the ignition timing on engines equipped with an exhaust emission-control system is performed in the same manner as setting the timing on conventional engines. Instructions for removing the distributor vacuum line and capping the manifold opening during timing setting, and the caution about using a timing light adapter rather than puncturing secondary cable insulation, also apply to engines with California exhaust control systems.

The important factors relative to setting ignition timing on engines equipped with exhaust control systems are that the correct timing specification be used and that the timing is accurately set.

Use Correct Timing Specifications

Usually the specifications for standard engines are given preference in the listing. California specifications may be covered with an asterisk or by footnotes. It is easy to misinterpret these specifications, so use care when selecting the timing specification for the engine being tuned.

Set the Timing Correctly

The ignition timing on many engines equipped with exhaust emission control systems is set at top dead center or after top dead center. This is contrary to conventional timing practice. The timing marks on some engines have been altered to indicate both before top dead center and after top dead center positions. Some timing marks are scribed with three letters, one above the other. They are A, O, and R. The A means advance or before top dead center position. The O stands for zero to top dead center position. The R means retard or after top dead center. Each line scribed on the marker usually stands for 2 degrees of adjustment; however, on some models it may be 4 or 5 degrees.

American Motors uses a similar marking in which the letter B signifies the before top dead center portion of the marker.

Chrysler Corporation cars are clearly maked with the words "Before" and "After" with a 0(zero) or top dead center position set between the before and after markings.

Ford Motor Company cars, equipped with thermactor emission control systems, are not timed later than top dead center, so the conventional Ford timing marks have been retained.

After setting the basic ignition timing, test the distributor centrifugal and vacuum advance mechanisms. Be sure to test both the spark advance and the spark retard action on dual-action vacuum units.

Engine Oscilloscope Testing

An engine analyzer or oscilloscope permits the service technician to actually look inside an engine operating under normal conditions. This is made possible by displaying an electrical pattern on a cathode tube and watching the electrons in action. Modern engine analyzers are equipped with not only the oscilloscope section for the display of patterns, but are also equipped with a voltmeter, a dwell and spark advance meter, and a tachometer. The voltmeter usually has two scales. One is for small-scale reading and one is for higher-voltage reading. This also holds for the tachometer. Most have a normal tachometer for higher rpm, as well as an expanded section for reading slight changes. Many engine analyzers are also equipped with an exhaust emission tester. This calls for two additional meters so that the carbon monoxide and hydrocarbon levels can be read. By using a modern engine analyzer, the technician can make a dynamic test of compression, ignition, and carburetion, and totally check out the cranking and starting conditions of spark ignition engines.

Leads and Connections

Oscilloscopes are equipped with various leads and connections so that readings can be made on the assigned instruments. These leads are color coded to insure proper hook-up. Prior to operating an oscilloscope, the service technician should be instructed in making the proper connections.

Before making the actual hook-up, the technician should first check to see the type of system used. For example, a leading ignition system is set up in a different way than a trailing ignition when working on a rotary engine. Some engines require special adapters to complete the hook-up. This is the case when working on solid-state systems.

As the technician is connecting an engine analyzer to a vehicle, he should make a visual check of all items.

Loose connections, excess dirt and grease, and visually damaged components of the vehicle should be corrected first.

© Copyright 1986, Tune-Up Manufacturers Institute

Testing Procedures

Once the visual check has been made and the engine analyzer has been hooked up correctly, the technician is ready for the testing procedures. As an engine analyzer is designed to follow a programmed procedure, the first test will show coil input voltage, cranking voltage, dwell, dwell variation, and point resistance. As this is part of the primary system, it will also show the primary pattern on the scope. The condenser and coil oscillations, as well as the closing and opening of the points in a conventional ignition system, can be seen visually by looking at the scope pattern. One should compare this pattern with the normal primary pattern shown in the equipment instructions.

The next test is the scondary ignition system. This will check coil polarity, spark line comparison, high-tension wiring, and coil secondary condition of plug wires.

During the secondary test position, it is possible to look at individual cylinder patterns by using the raster control on the tester. This spreads the cylinders so that possible problems can be found in any single cylinder.

The next test procedure will check the condition of the spark plugs, rotor, air gap voltage, and the unbalanced idle mixture.

The condition of the coil output, voltage input, and the spark plug firing voltage is the next programmed test.

After the above tests have been completed, the next test is to check the basic timing and the timing advance.

The last test is for carburetor balance, idle speed adjustment, idle mixture adjustment, cylinder balance, regulated voltage, and alternator condition.

If possible, a test should be made with the emission analyzer for CO and HC emissions.

An oscilloscope or engine analyzer gives the service technician the ability to totally analyze a vehicle's engine. This is accomplished by going through each of the programmed testing procedures and by noting the meter readings and analyzing the patterns shown on the cathode tube. There are many patterns showing different conditions or different components within the system. Because of this, the technician should only learn what the good patterns should look like until he has gained experience. If the pattern is not a good pattern on the scope, the technician should refer to the operating instructions and analyze the pattern for the problem.

Operating manuals and instructions are furnished with all engine analyzers. Because the modern engine analyzer is designed to do so many tests, it is necessary that the service technician take advantage of the instructions and operating manuals in detail. This will make it possible to understand all controls and functions, as well as to maintain the equipment.

Propane-Assisted Idle Adjustment Procedure

Many cars and light trucks with catalytic converters require a special idle mixture test and adjustment procedure to insure that the vehicle runs correctly and meets emissions standards. Prior to setting the idle mixture, the service technician should review the correct service manual for the vehicle's emission control information.

The catalytic converter contains a catalyst that reduces the carbon monoxide (CO) in the exhaust to a point that is nearly immeasurable at the exhaust pipe. Because of this, an infrared analyzer cannot accurately be used for the idle mixture adjustment. To correctly set the idle mixture adjustment, a metered amount of propane gas is injected into the carburetor and the rpm increase is noted while the engine is idling. This is called the *propane gain* or *aritificial enrichment test.* The idle mixture or aritificial enrichment specifications are usually listed on the *Vehicle Emission Control* decal found under the hood of the vehicle. They are also listed in the service manual for the vehicle. If, when the propane is injected into the carburetor, the gain is less than specified, the air-fuel ratio is too rich. If the gain is more than specified, the air-fuel ratio is too lean.

Before making any adjustments to idle fuel mixtures, the technician should test to see if adjustments are really needed. This is done by utilizing the following steps:

1. Connect an easy-to-read tachometer to the engine. It must have a large scale so that small increases or decreases in rpm will be visible.
2. Run the engine at a fast idle to stabilize the engine at operating temperature. This is a must.
3. The ignition timing must be set to specifications.
4. All necessary hoses must be removed and the holes plugged.
5. On those models so equipped, disconnect the thermactor air supply hose from the check valve.
6. Disconnect the evaporative emission purge hose from the air cleaner.
7. Check specifications for transmission position and place in either neutral or drive. If specifications call for the transmission to be in drive, always block the wheels and set the brakes.
8. A special tool is required for this step. This tool is an adjustable flow valve, hose, and special connector. This assembly is attached to an upright propane bottle, and the special connector is inserted into the air cleaner connection where the evaporation emission hose was previously attached.
9. The idle rpm will be adjusted next. The propane valve is to be slowly opened until the maximum idle speed is obtained. The maximum rpm gain should be noted and written down for reference. If the propane valve is opened beyond the maximum rpm gain, the rpm will drop off due to the mixture becoming too rich.
10. If the rpm increase is within the recommended specifications, the enrichment tool or propane adapter should be removed form the air cleaner and all hoses reconnected. If the rpm increase is to high or too low, the idle mixture must be corrected.

If idle adjustments are required, the following procedure will be used. When the idle speed rpm increase using propane is above specification, the gasoline mixture should be made richer by turning the mixture screws counterclockwise in equal amounts. (This is done without propane.) The screws should be turned until the rpm increases the same amount that the first reading was in excess of specifications. As an example, if the specifications call for an increase of 50 rpm and the observed reading was 70 rpm, the system must be enriched by 20 rpm. The adjustment screws must be turned until the rpm increases 20 rpm. This should then be rechecked using propane.

When the idle speed rpm increase with propane is below specifications, the mixture must be leaned. (This is also done without propane.) The mixture screws will be turned clockwise in equal amounts until the RPM drops the same amount that the original reading was below specifications. As an example, if the specifications call for an increase of 50 rpm and the observed reading was 30 rpm, the system must be leaned by 20 rpm. This should then be rechecked using propane.

There are times when the technician will find that the idle increase using propane is zero, and the minimum specified rpm or speed gain is zero. When this happens, the mixture should be made leaner by turning the limiter screws clockwise to the full rich stop position. The idle speed adjustment should then be set to specifications. The idle rpm increase should then be rechecked with propane. If the rpm gain is now greater than zero and within the specifications, the idle speed should be reset. If the speed gain is still zero, the mixture should be leaned by turning the limiter screws clockwise to their full position. The rpm drop should be noted during this step. If the rpm drop is equal to or greater than the manufacturer's specifications, the mixture screws should be returned to the original positions. If the rpm drop is lower than specifications, the mixture must be adjusted using the same steps that are used when the idle increase is below specifications. This calls for the mixture to be leaned by the required amount to meet specifications.

The special propane adapter tool is available from automotive distributors handling tools and equipment.

Hydrocarbon and Carbon Monoxide Exhaust Testing

Many types of exhaust gas analyzers are available on the market. Most service technicians are familiar with the *Wheatstone bridge analyzer,* which measures the air-fuel ratio on a percentage scale. This type of analyzer only measures the air-fuel ratio and does not measure hydrocarbons (HC) and carbon monoxide (CO). In those areas where exhaust emission standards are established, the test equipment used for exhaust analyzing must be able to test for hydrocarbon and carbon monoxide.

Hydrocarbon and carbon monoxide testers are available from any sources. They are known as *infrared equipment HC-CO analyzers,* or *exhaust emission testers.* It is the responsibility of the service technician to make sure he uses equipment that meets his area's requirements.

Most exhaust emission testers or infrared equipment use meters for the readings. Carbon monoxide is measured as the percentage of the exhuast, and hydrocarbons are measured as parts per million. This is accomplished by the tester making a quantitative analysis of the CO and HC present in the exhaust by measuring the amount of infrared energy absorbed by the exhaust gas sample.

The carbon monoxide (CO) meter indicates the richness or leanness of the fuel mixture. A high CO number indicates a rich mixture generally caused by one of the following:

1. Air cleaner (dirty or restricted).
2. Incorrect choke operation.

3. Idle speed not to specifications.
4. Incorrect carburetor float level.
5. Bad emission control air pump.
6. Poor idle mixture.

The hydrocarbon (HC) meter tells the service technician what is happening within the combustion chamber of the vehicle. If the gasoline is not burned in the chamber, raw fuel will exit out the exhaust pipe and will register on the HC meter. A high HC reading can be caused by one of the following:

1. Bad spark plugs.
2. Incorrect initial timing.
3. High-tension wires (plugs).
4. Exhaust and intake valves.
5. Poor rings or pistons.
6. Excessively lean or excessively rich idle mixture.
7. Air pump problem.

It is most important to follow the maintenance instructions that come with the equipment. Filters and other devices within the unit can absorb water and contaminants and cause the unit to give incorrect readings. Calibration is also very important, as this type of equipment is sensitive to altitude change. Calibration relative to regulations within the area or the state must always be checked.

An emission tester is an excellent tool to use in the diagnosis of vehicle problems.

Figure No. 191

SUPPLEMENTARY TUNE-UP SERVICES

1 – Inspect Drive Belts
2 – Check Manifold Heat Control Valve
3 – Service Carburetor Air Cleaner
4 – Check Fuel Filter
5 – Replace Positive Crankcase Ventilation System Valve
6 – Check Valve Clearances

© Copyright 1986, Tune-Up Manufacturers Institute

SUPPLEMENTARY TUNE-UP SERVICES

Several other units on the automobile engine have a definite effect on engine operation and the limiting of emissions. The full benefits of tune-up and emission control will be realized only when these units are also functioning properly.

The following items should be checked before your tune-up operation. If their service has been neglected during normal lubrication and service intervals, their condition will undesirably influence your tune-up.

1. Inspect condition and test tension of all drive belts. Slipping belts result in an undercharged battery, even though the charging system is functioning properly; engine overheating; loss of power steering assist; loss of air-conditioning efficiency; and lack of sufficient air supply from the air-injection reactor pump. A check of any of these systems *always* starts with an inspection and test of the drive belts.

2. Check manifold heat control valve action. A valve stuck in the open position will cause poor cold-engine operation and prolonged engine warm-up with extended use of the choke. A valve stuck in the closed position will result in loss of performance when normal operating temperature is reached. It can also result in burned engine valves due to an excessively lean carburetor fuel mixture.

3. Service or replace dirty air cleaners. A dirty carburetor air cleaner acts as a partial choke, upsetting the carburetor air-fuel ratio and making a fine carburetor adjustment impossible.

4. Check the fuel filter. A partially clogged fuel filter will restrict the fuel supply at high speed, reesulting in a performance complaint. A clogged fuel filter has the same effect on engine operation as a defective fuel pump.

5. Replace the positive crankcase ventilation system valve. Because most PCV-equipped engines are calibrated or adjusted to accommodate the additional air drawn from the crankcase, a clogged valve upsets the carburetor mixture, causing a rough idle and making a first-class tune-up impossible.

6. Check the valve clearance adjustment when necessary. Loose tappets are noisy. Tight tappets hold valves off their seats, causing engine roughness and resulting in burned-out valves. Valves should be adjusted when the need is indicated. Some late models have nonadjustable valves. If valves are noisy on these models, lifter must be replaced.

© Copyright 1986, Tune-Up Manufacturers Institute

Notes

ELECTRICAL AND ELECTRONIC SYMBOLS

AMPLIFIER (**Indicates Type of Amplifier)	Dual Input	
AUDIO SIGNALING DEVICES	Buzzer IEC	Bell IEC
CAPACITOR	CAP -1 (mfd or pfd value) IEC	
CIRCUIT BREAKER AND MANUAL STARTER	Magnetic IEC	Thermal
CONDUCTOR	Shielded Single Conductor	
CONNECTION	Connections / No Connection IEC	
CONNECTOR	Plug Contact Designation (A) (B) (C)	Plug Connector (male and female) IEC
FUSE	FU -1 (value) IEC	

	A.C.	D.C.
GENERATOR	(G with ~) IEC	(G) IEC
GROUND		⊥ IEC
INDUCTOR		⌒⌒⌒ IEC
JACK	Phone Jack — (2-circuit with break contact); (3-circuit with break contact)	Phone Plug — (2-circuit); (3-circuit)
MECHANICAL AND ELECTRICALLY RELATED DEVICES	(disengaged) (engaged) (Clutch IEC); Mechanical Linkage — — — — — IEC	Magnetically Dependent; Interlock — Mechanical --×--
METER	*Type may be indicated by abbreviation: A (Ammeter) IEC V (Voltmeter) IEC W (Wattmeter) IEC Tach (Tachometer) TT (Time meter) IEC	(*) IEC

RELAY (** Relay Designation)	Relay Contacts (normally open) IEC	Relay Contacts (normally closed) IEC
	Time Delay to Close TDC IEC	Time Delay to Open TDO IEC
	Time Delay to Open TDO IEC	Time Delay to Close TDC IEC
	Relay Coil IEC	Time Delay Relay
	Relay Contacts (transfer) IEC	
	Activating Device for Thermal Operated Relay HTR-1 IEC	Overload Relay Contacts (heat sensing)
RESISTOR AND RESISTANCE DEVICES	Adjustable Resistor RES -1 (value) INC (w/direction arrow)	Potentiometer POT -1 (value) INC (w/direction arrow)

© Copyright 1986, Tune-Up Manufacturers Institute

RESISTANCE DEVICES (continued)	Tapped Resistor RES -1 (value) — IEC	RESISTOR RES -1 (value) — IEC	
		Photoresistive Cell	
SEMICONDUCTOR DEVICES	Photoemmissive Diode — IEC	Photosensitive Diode — IEC	
		Zener Diode — IEC	
	Suppressor — IEC	Diode A — K (REC -1) — IEC	
	Silicon Controlled Rectifier (SCR) (REC -1)	Bridge Type — IEC	
	PNP Transistor — IEC	NPN Transistor — IEC	
	Unijunction Transistor — IEC		

© Copyright 1986, Tune-Up Manufacturers Institute

SOLENOID		IEC
SWITCH DEVICES	MOMENTARY CONTACT Single Pole Double Throw IEC	MAINTAINED CONTACT Single Pole Single Throw IEC
	PUSH SWITCH Normally Closed	Single Pole Double Throw IEC
	Normally Open	Three Position Center Off OFF IEC
	Two Circuit	Double Pole Double Throw
	ISOLATING FUSE-SWITCH IEC	Multiple Position IEC
	Normally Open	PRESSURE SWITCH Normally Closed
	Normally Closed THS -1 IEC	THERMOSTAT Normally Open THS -1 IEC
		MERCURY SWITCH Mounted for Break Operation

SWITCH DEVICES (Continued)	Selector Push Switch JOG REV. — FOR. A — B C - - - D | JOG ||	
Circuit	For. || Rev.	
	Free	Dep.	Free	Dep.
A - B	O	X	O	O
C - D	O	O	O	X | O = Open contacts X = Closed contacts	**PRECISION SWITCH** Normally Open Held Closed
		Normally Closed
		Normally Closed Held Open
		Normally Open
		Single Pole Double Throw
		Cam Operated (Indicate Degree of Actuation) 90°
TERMINAL CONNECTIONS	Plug-in Component Pin Connections ① ㊅⑥	Terminal Strip Connections TB -1
	□ ⬠ ⬡ △	Component terminal connections other than plug-in. Reference number or letter indicates marking on component.

© Copyright 1986, Tune-Up Manufacturers Institute

427

TRANSFORMER (* Voltage Designation)	Transformer IEC	Tapped Transformer (1 ph — 2 wind.) IEC
	Adjustable Autotransformer	(1 phase — 3 winding) IEC
	Fixed Autotransformer	Polyphase IEC
VISUAL SIGNALING DEVICES	Fluorescent Lamp FL	Neon Type Lamp
	* Lens color may be indicated by abbreviation A — Amber OP — Opalescent B — Blue P — Purple C — Clear R — Red G — Green W — White Y — Orange Y — Yellow	Filament Type Lamp

ABBREVIATIONS AND GLOSSARY OF TERMS

ABBREVIATIONS

Ford Motor Company

ACT	Air Charge Temperature
BMAP	Barometric and Manifold Absolute Pressure
BP	Barometric Pressure
CANP	Canister Purge
CFI	Central Fuel Injection
CP	Crankshaft Position
ECA	Electronic Control Assembly
ECT	Engine Coolant Temperature
EEC	Electronic Engine Control
EFI	Electronic Fuel Injection
EGO	Exhaust Gas Oxygen
EGR	Exhaust Gas Recirculation
EGRC	EGR Control
EGRV	EGR Vent
EVP	EGR Valve Position
FBC	Feedback Carburetor
FBCA	Feedback Carburetor Actuator
IAT	Intake Air Temperature
IM	Ignition Module
LOS	Limited Operation Strategy
MAP	Manifold Absolute Pressure
MCU	Microprocessor Control Unit
SSD	Sub System Diagnostic
TAB	Thermactor Air Bypass
TAD	Thermactor Air Diverter
TKA	Throttle Kicker Actuator
TKS	Throttle Kicker Solenoid
TP	Throttle Position
VAF	Vane Air Flow
VAT	Vane Air Temperature
VREF	Voltage Reference Signal
WOT	Wide Open Throttle

© Copyright 1986, Tune-Up Manufacturers Institute

General Motors

AIR	Air Injection Reaction
ALCL	Assembly Line Communications Link
ALDL	Assembly Line Data Link (Same as ALCL)
BARO	Barometric Pressure
CCC	Computer Command Control
CCP	Controlled Canister Purge
DFI	(DEFI) Digital Fuel Injection
DVOM	Digital Volt-Ohm Meter
EAC	Electric Air Control
EAS	Electric Air Switching
ECM	Electronic Control Module
EEC	Evaporative Emissions Control
EFI	Electronic Fuel Injection
EGR	Exhaust Gas Recirculation
ESC	Electronic Spark Control
EST	Electronic Spark Timing
HEI	High Energy Ignition
IAC	Idle Air Control
ISC	Idle Speed Control
MAP	Manifold Absolute Pressure
N.C.	Normally Closed
N.O.	Normally Open
O_2	Oxygen Sensor
PCV	Positive Crankcase Ventilation
P/N	Park/Neutral
PROM	Programmable Read Only Memory
TBI	Throttle Body Injection
TCC	Transmission Converter Clutch
THERMAC	Thermostatic Air Cleaner
TPS	Throttle Position Sensor
TVS	Temperature Vacuum Switch
VAC	Vacuum
VIN	Vehicle Identification Number
VSS	Vehicle Speed Sensor
WOT	Wide Open Throttle

© Copyright 1986, Tune-Up Manufacturers Institute

GLOSSARY OF TERMS

Air-fuel Ratio (A/F Ratio): The amount of fuel added to intake air as compared by weight.

Air Gap: Space between two electrical contacts.

Alternating Current (ac): A current passing through a series of positive and negative values to form a cycle.

Ambient: Surrounding atmosphere.

Ammeter: Instrument used to measure current flow.

Ampere: The practical unit of current measurement. Equal to the amount of electrical current required to produce 1 volt passing through a resistance of 1 ohm.

Ampere Hour Capacity: Term used to indicate the electrical capacity of a battery.

Analog Computer: A device that solves given mathematical problems by using electrical voltages as numerical variables in the problem. Output varies as a continuous function of the input.

Anode: Positive terminal.

Armature: A portion of an electrical device that moves within the flux of field windings.

Atomize: Reduction of gasoline into small droplets for better mixing with air.

Atmospheric Pressure: The weight of the earth's atmosphere, approximately 14.7 psi at sea level.

Base: That portion of a transistor used to control current flow through the emitter and collector.

Battery: A dc voltage source that converts chemical or thermal energy into electrical energy.

Bit: The smallest element of information in binary language.

Capacitor: A device constructed by separating two conductors with a dielectric. It has the ability to store an electrical charge in the dielectric.

Catalyst: A material that causes a chemical reaction between two compounds, without itself being affected.

Cathode: The negative terminal.

Chip: A silicone section containing all the elements of a complex electrical circuit.

Collector: One of the current-carrying poles of a transistor.

Computer: A device capable of receiving electrical information, applying programmed processes to the information, and issuing a result.

Conductor: Any material that can conduct electrical current.

Detonation: The sound created by excessive pressures in the combustion chamber.

Dielectric: A nonconductor, an insulator.

Digital Computer: A computer that receives information and issues commands in binary language form. Digital forms are on-off signals represented in square-wave pattern.

Diode: A device that has a high resistance to current flow in one direction and a low resistance in the opposite direction.

Direct Current (dc): An electrical current that flows in one direction.

Duty Cycle: The amount of time electrical current flows through a circuit as compared to the amount of time the circuit is open. A 50 percent duty cycle indicates that a circuit is live for the same length of time it is de-energized.

Dwell: The duty cycle represented in terms of angular degrees. Also referred to as *cam angle.*

Economizer: Sometimes called a *power valve.* Used to meter fuel discharge in relation to manifold vacuum.

Electrode: An electrical conductor within a device. The center core of a spark plug.

Electrolyte: Any substance that forms ions when placed in the proper environment. Forms a conductor of electricity.

Electromagnet: An iron core that is magnetized by electric current in an insulated coil surrounding the core.

Electromagnetic Interference (EMF): A magnetic distortion created by high-voltage carriers that interferes with electronic reception and transmission devices. Sometimes called radio-frequency interference (RFI).

Electromotive Force (EMF): A difference in electrical potential that causes the flow of current.

Electron: A particle fundamental to the makeup of an atom. It carries a negative charge of one electronic unit.

Emitter: A current-carrying pole of a transistor connected to the collector by controlling the base.

Field: The area affected by lines of magnetic force.

Float Level: The level of fuel in the bowl required to force the float arm against the carburetor needle and seat.

Flux: Electric or magnetic lines of force in magnetic field.

Frequency: The number of cycles or alternations occurring in 1 second. Measured in terms of cycles per second (cps) or more commonly hertz (Hz).

Fuse: An overcurrent protective device. A wire capable of carrying a specific maximum current. Exceeding the maximum causes the wire to thermally self-destruct.

Galvanometer: An instrument used to measure the amount and direction of electrical current in a circuit.

Grid: A metal meshlike screen used for electrical heating or conductance.

Ground: All modern vehicles use a negative ground electrical system. The negative terminal of a battery and electrical device connected to the chassis metalwork.

Growler: An instrument used to check armatures.

Hall Effect: The generation of a square-wave electrical signal, using a magnetic field to switch the base of a transistor.

Hardware: The physical components that comprise a computer system. Mechanical, magnetic, and electronic devices in a system.

Heat Riser: The flow of hot exhaust gases in a passage beneath the runners of an intake manifold.

Heat Sink: A metal mounting surface for electrical devices used to dissipate heat in the device.

Hertz (Hz): The number of cycles completed in 1 second.

Hydrocarbon: A substance composed of carbon and hydrogen (petroleum derivatives).

Hydrometer: A device for determining a battery's state of charge by measuring the specific gravity of the electrolyte.

Impedance: The total resistance of an electrical circuit or component. It is expressed in ohms.

Induction: The reaction of two different magnetic fields upon each other that are not electrically connected.

Injector: A solenoid or pop-off device that is used to supply metered amounts of fuel.

Insulator: A nonconductive material used to isolate an electrical circuit. The properties of an insulator are determined by its dielectric strength.

Integrated Circuit: The combining of several interconnected electrical circuits within the confines of a single block of material. An example is the imprinting of circuits onto a substrate of a silicon wafer.

Inverse Voltage: The voltage across a rectifier during the half-cycle when current does not flow.

Light-emitting Diode (LED): A diode constructed in such a way as to generate a source when biased in a forward direction.

Load: The amount of energy demand placed on an engine with a throttle valve. Manifold vacuum losses determine the extent of load.

Logic (Computer): A system of processing information where each function is affected by the preceding function.

Magnet (Permanent): A ferrite material charged to maintain a constant magnetic field.

Magnetostrictive: The change in resistance of a conductor as caused by application of magnetism.

Memory: That portion of a computer used to store information for retrieval at a later date.

Microfarad: The practical unit of measurement for capacitance.

Microprocessor: The basic form of a computer used to perform logic functions.

Milli: Prefix for indicating one-thousandth of a specified unit. *Example:* 20 millivolts = 0.020 volts.

Modulate: A constantly changing control. Reacting to opposing influences. *Example:* The opening and closing of a backpressure transduced EGR valve.

Monolithic: A structure composed of a single rigid substance. *Example:* The catalytic element of a monolithic converter.

Multiplexing: The sending or receiving of several electrical or light frequency signals through a single conductor.

Nitrogen Oxides (NO_x or NO_m): The combining of nitrogen with free oxygen. A pollutant generated by excessive combustion chamber temperatures. A major contributor to smog.

Noise: Electrical disturbance usually caused by excessive electromagnetic interference.

Normally Open or **Closed:** Describes the conductive position of a switch or relay while in the rest or de-energized position.

© Copyright 1986, Tune-Up Manufacturers Institute

NPN Transistor: A transistor constructed with a *P*-type base and *N*-type collector and emitter.

N-Type Material: A semiconductor material with impurities added to give it a majority of electrons as charge carriers.

Ohm: The unit of measurement for electrical resistance.

Open Circuit: A circuit not capable of conducting current flow.

Oscilloscope: A device containing a viewing screen that displays electrical output in the form of quantity (voltage) and time.

Oxidize: To combine with oxygen. *Example:* CO combined with O yields CO_2.

Piezoelectric: A crystal-type device that generates a small voltage when placed under mechanical stress. A detonation sensor can be piezoelectric.

PNP Transistor: A transistor constructed with an *N*-type base and *P*-type collector and emitter.

Polarity: The arrangement of north and south poles in a magnetic field. Positive and negative direction of electrical current flow.

Port: Cylinder head passage for intake or exhaust charges.

Potential: The difference in voltage between two points in a circuit.

Potential Drop: The loss of voltage due to resistance between two points in a circuit.

Preignition: Combustion being initiated prior to predetermined cylinder pressure levels.

Primary Circuit: Low tension wiring used to supply current to ignition coil.

Prom: Programmable read only memory. A section of the computer used to contain information for input and output control. Generally used to tailor a computer to a particular model application. A calibration assembly.

P-Type Material: A semiconductor material with impurities added to produce free holes in the material.

Rectifier: A device that has the ability to convert ac to dc.

Regulator: Component capable of controlling pressure. The pressure can be hydraulic, fuel, or electrical.

Relay: A device used to open or close an electrical circuit. Allows the switching of a high-current circuit with a low-current signal. A transistor is similar in operation to a relay.

Resistance: The opposition to electrical flow in a circuit or component.

Retard: The moving of ignition spark timing close to or after piston top dead center (TDC).

RFI: Radio-frequency interference. See *electromagnetic interference.*

Rise Time: The amount of time for coil saturation to reach 90 percent of potential.

Saturation: A circuit condition that does not change output when input is increased.

Secondary Voltage: The high tension delivered to the spark plugs when current flow in the primary circuit is interrupted.

Semiconductor: A device that can operate as a conductor or nonconductor, depending on the polarity of the applied voltage. A rectifier.

Short Circuit: The unwanted grounding of a conductor.

Signal: An electrical, luminescent, physical, or audible indication used to convey information.

Software: The language and procedures used to program a computer.

Solenoid: An electromagnetic device used to control electrical, pressure, or mechanical operation.

Solid State: Components constructed with semiconductors.

Spark Advance: The positioning of ignition timing in relation to crankshaft position.

Specific Gravity: Relative weight of substance as compared to the weight of water.

Stator: Windings that create alternating current in relation to armature (reluctor) movement.

Substrate: The surface on which integrated circuits are applied.

Tach Terminal: Negative side of coil used to indicate rpm.

Thermactor: Ford air-injection systems.

Thermal resistor: A device that changes resistance in relation to temperature. Resistance increases as temperature rises.

Thermistor: A solid-state device that changes resistance in relation to temperature. Resistance decreases as temperature increases.

Thick Film: The method of applying integrated circuits to a ceramic substrate with a silk-screen process.

Transducer: A device that receives energy from one system and transmits it to another system. The transmission may or may not be in a different form. *Example:* A manifold pressure sensor converts vacuum into an electrical signal and sends the signal to a computer system.

Transient Voltage: Voltage levels that exceed the maximum capacity of a circuit. Often called a *spike,* it can easily destroy solid-state components.

Transistor: A small semiconductor device originally designed to replace the vacuum tube. Its operation is similar to that of a relay.

Vacuum: Any pressure less than the ambient atmospheric pressure. Negative pressure.

Vapor Lock: The percolation (boiling) of fuel that causes gaseous vapors to affect the development of pressure in the fuel system.

Venturi: An aerodynamic restriction that creates a surface vacuum zone relative to the velocity of air passing over the restriction.

Volatility: The ability of a liquid to enter a gaseous state. Evaporative property.

Volt: A unit of electrical force required to cause a current of 1 ampere to flow through a resistance of 1 ohm.

Voltage Regulator: A device used to control charging system voltage output.

Voltmeter: An instrument used to measure electrical pressure (voltage) in a circuit.

Watt: The amount of electrical power needed to do work at a rate of 1 joule/second. Watts = amperes × volts.

Wave Form: Describes a continuous signal type. Square-wave forms are digital. AC waves are analog.

Zener Diode: A diode that conducts when reverse voltage reaches a predetermined value.

INDEX

A

Accelerator pump, 247
Air injection systems:
    air management systems, 271
    air pump, 269
    aspirator systems, 271
    check valve, 270
    diverter valve, 269
    operation, 269
Air management systems, 271
ALCL, 369
Alternator components:
    application, 75
    diode trio, 81
Alternator operating principles:
    full-wave bridge, 73
    theory, 73
    three-phase connection, 73
Alternator testing:
    Chrysler, 107
    Delcotron, 101
    with electronic regulators, 99
    factors, 115
    Ford, 113
    with integral regulator, 105
    with mechanical regulators, 97
Ammeter, 83
Ampere, 13
Ampere-turns, 23
Armature, 57
Available voltage, 159

B

Battery:
    amp-hour method, 51
    capacity test, 49
    cold cranking performance, 51
    dry-charge, 41
    fast charge test, 50
    leakage test, 43
    load test, 49
    maintenance-free, 49
    rating methods, 51
    reserve capacity, 52
    specific grafity, 45
    storage, 41
    three minute charge test, 49
Bendix drive, 61
Buildup time, 153

C

Capacitive discharge ignition, 181
Carburetor:
    accelerating system, 247
    adjustment, 259
    choke, 249
    electronic, 257
    float system, 237
    idle circuit, 239
    main metering circuit, 241
    metering rod, 241, 243
    power system, 245
Catalytic converters:
    dual stage, 334
    oxidation type, 333
    testing, 334
    three-way type, 333
Centrifugal advance, 149
Charging voltage test, 401
Check engine light, 369
Choke:
    application, 249
    automatic, 251
    diaphragm-type pulloff, 252
    electric assist, 255
    unloader, 251
    vacuum piston, 251
Chrysler EFC:
    inputs, 381
    outputs, 382
Chrysler electronic ignition:
    breakerless, 185
    lean burn, 187
    troubleshooting, 191
Circuit, electrical, 13, 17
Circuit testing tools:
    ammeter, 31
    jumper wire, 31
    ohmmeter, 32
    self-powered test light, 31
    test light, 31
    voltmeter, 32
Closed loop, 257
Commutator, 131
Compression test, 405
Computer Command Control (CCC), 363
Computerized EGR systems, 299
Computerized engine controls:
    Chrysler EFC, 381
    ECM, 363
    Ford EEC, 373

Computerized engine controls (*cont.*)
   G.M. diagnostic codes, 369
   inputs, 364
   introduction, 361
   outputs, 366
   sensors, 365
   solenoids, 367
Condenser, 161, 163
Conductor, 13
Constant voltage charging system, 71
Cranking voltage test, 399

D

Decel valve, 287
Detonation sensors:
   operation, 217
   servicing, 218
   testing, 218
Diodes, 35, 77
Diode tests, 119, 121
Distributor:
   cap, 167
   rotor, 167
Duty cycle, 257
Dwell angle, 143

E

Early fuel evaporation (EFE), 321
Electrolyte, 41
Electromagnetic fields, 23
Electronic control assembly (ECA), 373
Electronic control module (ECM), 363
Electronic engine control (EEC):
   description, 373
   inputs, 374
   outputs, 375
   self-test, 379
Electronic fuel injection (EI):
   air-delivery, 391
   electronic control network, 389
   fuel-delivery, 390
   introduction, 387
   multipoint, 389
Emission controls:
   air injection, 269
   assist units, 345
   catalytic converters, 333
   computerized EGR, 299
   crankcase, 263

   decel valve, 287
   EFE, 321
   EGR systems, 289
   electronic EFE, 323
   evaporative, 264
   exhaust, 263
   fuel evaporation, 313
   heated air systems, 273
   idle stop solenoids, 281
   introduction, 263
   late-model purge, 317
   modification type, 267
   oxygen sensor, 337
   PCV, 325
   ported vacuum, 310
   temperature vacuum switch, 311
Exhaust gas recirculation (EGR):
   application, 289
   back-pressure transducer, 290, 295
   checking, 290
   computerized testing, 309
   dual diaphragm, 293
   floor jet, 290
   Ford late model, 305
   General Motors solenoids, 301
   negative backpressure, 296
   positive backpressure, 295
   remote sensor type, 295
   vacuum amplifier, 289

F

Field, magnetic, 21
Field windings, 57
Field winding tests, 123
Firing order, 173
Flux lines, 21
Ford ignition systems:
   dual-mode timing, 211
   Dura-spark I, II, 201
   thick film integrated (TFI), 213
   troubleshooting Dura-spark I, II, 203
Forward-biased, 35
Four-stroke cycle, 9
Fuel evaporation systems:
   carb vapor control, 314
   charcoal canister, 313
   operation, 313
Fuel pump:
   action, 231
   electric, 233

in-line, 227
interruptor, 235
Fusible link and wires, 25

G

General Motors (Delco) ignition:
   electronic spark timing (EST), 221
   HEI, 195
   troubleshooting HEI, 197
Generator operating principles, 131
Generator testing, 133
Ground circuit, 27

H

Hall-effect switch, 215
Heat riser:
   EFE, 321
   electric EFE, 323
   thermal, 321
Heat sink, 77
Heated carburetor air systems:
   operation, 273
   testing, 277
High amperage circuit, 55
Hot idle compensator, 285
Hydrometer, 45, 47

I

Idle limiters, 355
Idle speed control (ISC), 367
Idle stop solenoid, 281
Ignition:
   breaker-point type, 137
   centrifugal advance, 149
   coil, 153
   detonation sensors, 217
   distributor assembly, 141
   Hall-effect, 215
   primary circuit, 139
   secondary circuit, 153
   spark advance, 147, 151
Indicator lamp, 85, 87
Insulator, 13

L

Lambda probe, 337
Lean-roll (stumble), 355

Loop:
   closed, 338
   definition, 337
   open, 337

M

Magnetism, theory, 21
Manifold pressure sensor (map), 389
Mixture-control solenoid, 257
Multipoint injection, 389

O

Ohm, 13
Ohm's law, 15
Open loop, 257
Oxide of nitrogen, 289
Oxygen sensor:
   M/C solenoid, 257
   operation, 337
   testing, 338

P

Polarity, 157
Ported vacuum, 310
Positive crankcase ventilation:
   operation, 325
   testing, 329
Power loss, 403
Prestolite (BID) ignition, 207
PROM, 364
Purge controls:
   bowl vent, 317
   computer control, 317

R

Reactance, 153
Regulators:
   fuel, 390
   voltage, 89–95
Resistor:
   ballast, 137, 139
   theory, 13, 67

© Copyright 1986, Tune-Up Manufacturers Institute

S

Secondary circuit:
  description, 153
  resistor spark plugs, 169
  RFI, 169
  suppression, 169
Sensors:
  pressure, 365
  throttle position (TPS), 365
  VSS, 365
Solid-state electronics, 35
Spark advance, 147
Spark delay valve, 343
Spark plug:
  features, 177
  hcat range, 175
  reach, 175
  resistor type, 169
  service, 409
Square wave, 215
Starting circuit, 55
Starting drives:
  inertial engagement, 59
  operation, 59
  overrunning-clutch-type, 59
Starting motor:
  amperage draw test, 63
  battery starter tester, 63
  ground circuit test, 67
  insulated circuit test, 65
  operation, 57
Starting switch, 61
Stator winding tests, 125
Stoichiometric, 337

T

Temperature vacuum switch:
  operation, 310
  testing, 349
Testing:
  diodes, 35
  essentials, 7
Throttle body fuel injection (TBI), 393
Timing, 411
Transistor, 36, 79
Transmission converter clutch (TCC), 367
Transmission regulated spark system, 341
Trickle charge, 71
Tune-up procedure, 397

V

Volt, 13
Voltage drop, 67
Voltage regulator, 89–95

W

Watt, 13

Z

Zener diodes, 36